How Language Began

ALSO BY DANIEL L. EVERETT

Don't Sleep, There Are Snakes:
Life and Language in the Amazonian Jungle

Language: The Cultural Tool

Dark Matter of the Mind

How Language Began

*The Story of Humanity's
Greatest Invention*

Daniel L. Everett

LIVERIGHT PUBLISHING CORPORATION
A Division of W. W. Norton & Company
Independent Publishers Since 1923
New York · London

Copyright © 2017 by Daniel Everett
First American Edition 2017

For information about special discounts for bulk purchases, please contact
W. W. Norton Special Sales at specialsales@wwnorton.com or 800-233-4830

Manufacturing by LSC Communications, Harrisonburg, VA
Production manager: Anna Oler

Library of Congress Cataloging-in-Publication Data

Names: Everett, Daniel Leonard, author.
Title: How language began : the story of humanity's greatest invention / Daniel L. Everett.
Description: First American edition 2017. | New York : Liveright Publishing
Corporation, a division of W. W. Norton & Company, [2017] |
Includes bibliographical references and index.
Identifiers: LCCN 2017025673 | ISBN 9780871407955 (hardcover)
Subjects: LCSH: Language and languages—Origin. | Human
communication. | Semiotics. | Psycholinguistics.
Classification: LCC P116 .E73 2017 | DDC 401—dc23
LC record available at https://lccn.loc.gov/2017025673

Liveright Publishing Corporation
500 Fifth Avenue, New York, N.Y. 10110
www.wwnorton.com

W. W. Norton & Company Ltd.
15 Carlisle Street, London W1D 3BS

1 2 3 4 5 6 7 8 9 0

Language is not an instinct, based on genetically transmitted knowledge coded in a discrete cortical 'language organ'. Instead it is a learned skill ... that is distributed over many parts of the human brain.

Philip Lieberman

For John Davey,
mentor and friend

Contents

Part Four: Cultural Evolution of Language

List of Figures

Picture Credits

Figure 5: Copyright © John Gurche; Figures 6, 13, 14, 15: Didier Descouens (CC-BY-SA-4.0) – Museum of Toulouse; Figures 9, 10: Copyright © Robert G. Bednarik; Figure 11: Human Origins Program, Smithsonian Institution; Figure 12: Wim Lustenhouwer, VU University Amsterdam; Figures 17, 18, 19: From Blumenfeld: *Neuroanatomy through Clinical Cases*, Second Edition, Sinauer Associates, Inc., 2010; Figure 20: Reprinted from *Neuroanatomy of Language Regions of the Human Brain*, Michael Petrides, Cytoarchitecture, Pages 89–138, Copyright 2014, with permission from Elsevier; Figure 21: www.theodora.com/anatomy, used with permission; Figure 22: http://www.internationalphoneticassociation .org/content/ipa-chart, available under a Creative Commons Attribution-Sharealike 3.0 Unported License. Copyright © 2015 International Phonetic Association; Figure 33: Figure 4.2.3, *Gesture and Thought*, David McNeill, 2005, University of Chicago Press.

Acknowledgements

A BOOK LIKE THIS depends on the help of many people, from those whose work I have made use of to those who actually took the time to read the manuscript and offer comments. It is a pleasure to thank Gisbert Fanselow, Vyvyan Evans, Caleb Everett, Peter J. Richerson, Helen Tager-Flusberg, Geoffrey Pullum and Philip Lieberman for helpful, often highly critical comments on portions of this manuscript. I especially want to thank Maggie Tallerman, for reading and commenting on just about every page. None of these readers agrees with me completely – some very little, in fact. I therefore absolve them of all responsibility for what follows, only thanking them for invaluable help. A special thanks goes to my editor at Profile Books, John Davey, for helping me shape this book in my head several years ago. Phil Marino and Bob Weil at Liveright provided me with the most helpful and detailed editorial suggestions that I have ever received. If this book is a success in any sense of that word, Phil, Bob and John are a large part of the reason. As always, it is a pleasure to thank Max Brockman, who has been a supporter and wonderful agent for years.

I thank Kristen Nill, my indefatigable and ever-thoughtful assistant for securing permissions for the illustrations in this book and for maintaining my calendar.

This work and much else are only possible because of the support and love of Linda Wulfman Everett.

Preface

AROUND 1920 A RATTLESNAKE killed my great-grandfather outside of Lubbock, Texas. Walking home from church with his family across a cotton field, Great-grandfather Dungan was telling his children to watch out for snakes in the field when he was suddenly struck in the thigh. His daughter, Clara Belle, my grandmother, told me that he suffered for three days, crippled in pain and screaming, until he finally expired in his bedroom at the back of the house.

One did not have to be at the scene of the incident to know that, because it was a rattlesnake, it must have 'warned' my great-grandfather before striking. But, considering the outcome, there must have been a communication failure between Papa Dungan and the snake. My grandmother saw the snake bite her father and she talked about the event a great deal during my childhood. She often remembered the moments when the snake was 'warning' her father, as if the beast would use actual words if it only could. However, people who know that rattlesnakes communicate often confuse their tail shaking with language, leading them to anthropomorphise and evoke human terms, such as saying they 'tell' threatening creatures 'to stay away' by shaking the keratin-formed, interlocking, hollow parts at the end of their tail to produce a loud rattle. Though that action is not technically language, the snake's rattling carries important information nonetheless. My great-grandfather paid a heavy price for failing to heed that message.

Rattlesnakes aren't the only animal communicators, of course. In fact, all animals communicate, receiving and transmitting information to other animals, whether of their own or different species. As I will later explain, however, we should resist labelling the rattle of a snake 'language'. A rattlesnake's repertoire is splendidly effective, but for severely limited purposes. No snake can tell you what it wants to do tomorrow or how it feels about the weather. Messages like those require language, the most advanced form of communication earth has yet produced.

The story of how humans came to have language is a mostly untold one, full of invention and discovery, and the conclusions that I come to through that story have a long pedigree in the sciences related to language evolution – anthropology, linguistics, cognitive science, palaeoneurology, archaeology, biology, neuroscience and primatology. Like any scientist, however, my interpretations are informed by my background, which in this case are my forty years of field research on languages and cultures of North, Central and South America, especially with hunter-gatherers of the Brazilian Amazon. As in my latest monograph on the intersection of psychology and culture, *Dark Matter of the Mind: The Culturally Articulated Unconscious*, I deny here that language is an instinct of any kind, as I also deny that it is innate, or inborn.

As far back as the work of psychologist Kurt Goldstein in the early twentieth century, researchers have denied that there are language-exclusive cognitive disorders. The absence of such disorders would seem to suggest that language emerges from the individual and not merely from language-specific regions of the brain. And this in turn supports the claim that language is not a relatively recent development, say 50–100,000 years old, possessed exclusively by *Homo sapiens*. My research suggests that language began with *Homo erectus* more than one million years ago, and has existed for 60,000 generations.

As such, the hero of this story is *Homo erectus*, upright man, the most intelligent creature that had ever existed until that time. *Erectus* was the pioneer of language, culture, human migration and adventure. Around three-quarters of a million years before *Homo erectus* transmogrified into *Homo sapiens*, their communities sailed almost two hundred miles (320 kilometres) across open ocean and walked nearly the entire world.

Erectus communities invented symbols and language, the sort that wouldn't seem out of place today. Although their languages differed from modern languages in the quantity of their grammatical tools, they were human languages. Of course, as generations came and went, *Homo sapiens* unsurprisingly improved on what *erectus* had done, but there are languages still spoken today that are reminiscent of the first ever spoken, and they are not inferior to other modern languages.

The Latin word *Homo* means 'man'. Therefore, any creature of the *Homo* genus is a human being. In two-word Latin biological nomenclature, a genus is the broader classification of which a species is a variant. Thus, *Homo erectus* describes a species – *erectus*, 'standing' – that is a member of the human, *Homo*, genus. *Homo erectus* thus means 'standing man'. This is the first species of humans. *Homo neanderthalensis* means 'Neander Valley man', based on the fact that its fossils were first discovered in the Neander Valley of Germany. *Homo sapiens* means 'wise man', and suggests, erroneously as we see, that modern humans (we are all *Homo sapiens*) are the only wise or intelligent humans. We are almost certainly the smartest. But we are not the only smart humans who ever lived.

Erectus also invented the other pillar of human cognition: culture. Who we are today was partially forged by the intelligence, travels, trials and strength of *Homo erectus*. This is worth stating because too many *sapiens* fail to reflect on the importance of earlier humans to who we are today.

My interest in language and its evolution is personal. All of my life, from my earliest years growing up on the Mexico–California border, languages and cultures have fascinated me. And how could they not? Incredibly, all languages share at least some grammatical characteristics, whether it be words for things, words for events or conventions for ordering and structuring sounds and words, or organising paragraphs, stories and conversations. But languages are perhaps even more unlike one another than alike. However easy or difficult these differences may be to discover, they are always there. Today, there is no universal human language, whether or not there was at some period in the remote past. And there is no mental template for grammar that humans are born with. Languages' similarities are not rooted in a special genetics for language. They follow from culture and common information-processing solutions and have their own individual evolutionary stories.

But each language satisfies the human need to communicate. While many people in today's world are tempted to spend more time on social media than perhaps they should, it is the pull of linguistic intercourse that is mainly driving them there. No matter how busy some are, it is hard for them to avoid entering into some 'conversation' on the screen in front of them to opine on issues about which they often know little and care less. Whether via water cooler conversations, or absorbing information from television, or discussing plays, or reading or writing novels, talking and writing bind humans ever more tightly into a community.

As a result, language – not communication – is the dividing line between humans and other animals. Yet it is impossible to understand language without understanding something of its origin and evolution. For centuries people have offered ideas about where and when language originated. They have wondered which of the many species of the genus *Homo* was the first to have language. And they have asked what language sounded like at the aurora of human history. The answer is easy. Language gradually emerged from a culture, formed by people who communicated with one another via human brains. *Language is the handmaiden of culture.*

How Language Began offers a unique, wide-ranging story of the evolutionary history of language as a human invention – from the emergence of our species to the more than 7,000 languages spoken today. Their complexity and range was invented by our species, later developing into local variants, each new linguistic community altering language to fit its own culture. To be sure, the first languages were also constrained by human neurophysiology and the human vocal apparatus. And all languages came about gradually. Language did not begin with gestures, nor with singing, nor with imitations of animal sounds. Languages began via culturally invented symbols. Humans ordered these initial symbols and formed larger symbols from them. At the same time symbols were accompanied by gestures and pitch modulation of the voice: intonation. Gestures and intonation function together and separately to draw attention to, to render more salient perceptually, some of the symbols in an utterance – the most newsworthy for the hearer. This system of symbols, ordering, gestures and intonation emerged synergistically, each component adding something that led to something more intricate, more effective. No single one of these components was part of language until they all were – the whole giving purpose to the parts – as far back as nearly two million years ago. Language was culturally invented and shaped and made possible by our large, dense brains.[1] This combination of brain and culture explains why only humans have ever been able to talk.

Other authors have labelled language an 'invention', only to qualify that reasonable assessment by adding 'but it's not *really* an invention. That is a metaphor.' But the use of the word 'invention' here is not a metaphor. It means what it says – human communities *created* symbols, grammar and language where there had been none before.

But what is an invention? It is a *creation of culture*. Edison alone did not invent the light bulb; he needed Franklin's work in electricity nearly 200 years before him. No one person invents anything. Everyone is part of a culture and part of each other's creativity, ideas, earlier attempts and the general world of knowledge in which they live. Every invention is built up over time, bit by bit. Language is no exception.

Introduction

In the beginning was the Word.

John 1:1

No, it wasn't.

Dan Everett

IT WAS A SULTRY MORNING IN 1991, along the Kitiá river in the Amazonian rainforest of Brazil, some 200 miles (320 kilometres) in a single-engined plane from the nearest town. I found myself fitting headphone mics on two slender, weather-hardened men, Sabatão (sa-ba-TOWN) and Bidu (bee-DOO). This time of day would usually find them in the jungle, armed with eight-foot blowguns and quivers of poisoned darts, hunting for peccary, deer, monkey, or other game indigenous to their Eden. But today they were going to talk to each other while I bothered with recorder controls and sound levels.

Before we began I explained to them, again, in a mix of their language, Banawá (ba-na-WA) and Portuguese, what I wanted. 'Talk to each other. About anything. Tell each other stories. Talk about the Americans and the Brazilians who visit the village. Whatever you want.' I had coaxed and paid them to be here because I was after the holy grail of the linguistics field researcher – natural conversation (interactive, spontaneous communication involving more than one person). I knew from my past failures that natural conversations were nearly impossible to record. This is because the presence of a field researcher with recording equipment affects the perception of the task and contaminates the result so severely that one usually gets only stilted, unnatural exchanges that no native speaker would accept as a real conversation. (Imagine if someone sat you down with a friend, fitted you with a headset mic and then cued you, 'Converse!')

But today, as I tested the sound quality of the recording I was making, I could barely contain my excitement. They began like this:

Sabatão: Bidu, Bidu! Let talk today.
Bidu: *Mmm.*
S: Let talk in our language.
B: *Mmm.*
S: Daniel likes our language very much.
B: *Yes, I know.*
S: I will talk. You can then tell a story about that jaguar.
B: *Yes.*
S: Let's remember how things were a long time ago.
B: *Yes. I remember.*
S: A long time ago the whites arrived. A long time ago
            the whites arrived in our village.
B: *Them I know.*
S: They found us. We will work with them.
B: *Yes. Them I know.*

Their conversation glided from topic to topic naturally for the better part of an hour.

Though I was several thousand miles from home, sweating profusely, swatting away wasps and blood-sucking flies, I nearly teared up after Sabatão and Bidu finished, forty-five minutes later. I thanked them enthusiastically for this verbal treasure they had provided me with. They smiled and left to go hunting with their blowguns and poisoned darts. I continued alone, transcribing (writing down every nuance phonetically), translating and analysing the recording. After a couple of days of hard work to make the data 'presentable', I turned over the recordings, my notes and the bulk of the remaining work of analysis to a graduate student who had accompanied me to the Amazon from the University of Manchester in England.

At the end of the day our research team – myself and three students – enjoyed an evening meal of beans, rice and peccary meat I had purchased from the Banawás. We sat around after the meal, discussing the jungle heat and bugs, the likes of which we'd never seen before, but especially we conversed about the recorded conversation of Bidu and Sabatão and how grateful we were to them. Conversations within conversations. Conversations about conversations.

Following the blink-of-an-eye Amazonian sunset, the Banawás came to visit, as is their custom. The four of us made Kool-Aid and coffee and opened a package of sweet biscuits for them. We first greeted the Banawá women. The female students handled most of the serving and greeting of the women as is culturally appropriate among the Banawás, who practise rigorous segregation of the sexes. Soon the men were allowed to sit down and we served more coffee, Kool-Aid and sweet biscuits. As we ate and drank, we chatted with the men, mainly answering their questions about our families and homes. Just like people everywhere do on a daily basis, we and the Banawás were building relationships and friendships through conversation.

Natural conversations of this sort are important to linguists, psychologists, sociologists, anthropologists and philosophers because they embody the complex, integrated whole of language in a way that no other manifestation of language does. Conversations are the apex of linguistic studies and sources of insight particularly because they are potentially open-ended in meaning and form. They are also crucial to understanding the nature of language because of their 'underdeterminacy' – saying less than what is intended to be communicated and leaving the unspoken assumptions to be figured out by the hearer in some way. Underdeterminacy has always been part of language.

As an example of underdeterminacy, look at the second line of the conversation between Sabatão and Bidu. Sabatão says to Bidu, 'Let talk in *our language.*' This is strange if one takes it literally, because *they are already speaking in their language.* In fact, these men would be hard pressed to carry on a natural conversation in Portuguese, because their knowledge of it is rudimentary, limited principally to bartering. Sabatão's words presume something unstated. Sabatão is using these words to indirectly let *me* know that they will not use any Portuguese to converse *because they know that I am trying to understand how they converse in their language* and *because they want to help me.* None of this is spoken. Though underdetermined by the words, it is implicit in the context.

Likewise, in the line 'Let's remember how things were a long time ago' there is shared knowledge of the general range of things they are trying to remember. What is up for grabs here? Rituals? Hunting? Relationships with other peoples? How long ago? Before the Americans came? Before the Brazilians came? A hundred generations? Both Bidu and Sabatão (or indeed any other Banawá) know what is being talked about, but this is not clear initially to someone from another culture.

Sabatão and Bidu are two of the eighty or so remaining speakers of Banawá, a language that has already helped the scientific community learn a great deal about human language, cognition, the Amazon and culture. Specifically, they have taught us about unusual sound structures and grammar, the ingredients and process for manufacturing poison for darts and arrows, their classification of Amazonian flora and fauna and their connections linguistically to other Amazonians. Such lessons naturally follow from working out the knowledge structures, values, linguistics and social organisation of different groups who, like the Banawás, have spent millennia mastering life in a particular niche.

Any community – whether it be the Banawás, the French, the Chinese, or Botswanans – uses language to build social ties between members of their community and others. Indeed, our species has been conversing for a very long time. All languages on earth trace their underdetermined, socially bonding, grammar-constrained, meaning-motivated expressions of thought back to early hominins, to *Homo erectus* and perhaps even earlier. Based on the evidence of *Homo erectus* culture, such as their tools, houses, village spatial organisation and ocean travel to imagined lands beyond the horizon, the genus *Homo* has been talking for some 60,000 generations – quite possibly more than one and a half million years. By now one would expect our species, after more than a thousand thousand years of practice, to be very good at language. And we would also expect the languages we have all developed over time to better fit our cognitive and perceptual limitations, auditory range, vocal apparatus and brain structures. Underdeterminacy means that every utterance in every conversation and every line in every novel and each sentence of any speech contains 'blank spots' – unspoken, assumed knowledge, values, roles and emotions – underdetermined content that I label 'dark matter'. Language can never be understood entirely without a shared, internalised set of values, social structures and knowledge relationships. In these shared cultural and psychological components, language filters what is communicated, guiding a hearer's interpretations of what another is saying. People use the context and cultures in which they hear language to interpret it. They also use gestures and intonation, in order to interpret the full meaning of what is being communicated.

Like all humans, the first *Homo* species to begin the long arduous process of constructing a language from scratch almost certainly never said entirely what was on their minds. That would violate basic design features of language. At the same time, these primordial hominins would not have simply made random sounds or gestures. Instead, they would have used means to communicate that they believed others would understand. And they also thought their hearers could 'fill in the gaps', and connect their knowledge of their culture and the world to interpret what was uttered.

These are some of the reasons why the origins of human language cannot be effectively discussed unless conversation is placed at the top of the list of things to understand. Every aspect of human language has evolved, as have components of the human brain and body, to engage in conversation and social life. Language did not fully begin when the first hominid uttered the first word or sentence. It began in earnest only with the first conversation, which is both the source and the goal of language. Indeed, language changes lives. It builds society, expresses our highest aspirations, our basest thoughts, our emotions and our philosophies of life. But all language is ultimately at the service of human interaction. Other components of language – things like grammar and stories – are secondary to conversation.

This point raises an interesting question about language evolution, namely who spoke first? Over the past two centuries a plethora of ancestors for humans have been proposed, from South Africa, Java and Beijing, to the Neanderthal Valley and Olduvai Gorge. At the same time, researchers have proposed several novel hominin species, leading to a confusing evolutionary mosaic. To avoid getting caught up in a morass of uncertain proposals, only three language-possessing species need to be discussed – *Homo erectus*, *Homo neanderthalensis* and *Homo sapiens*.

Few linguists claim that *Homo erectus* had language. Many, in fact, deny this. There is currently no consensus on when the first humans spoke. But there does seem to be some modern consensus on human evolution, the methods used and an overview of the evolution of our species' physical and cognitive abilities. In *The Descent of Man*, Charles Darwin suggested that Africa might be the birthplace of humans because it is also the location of most apes. He reasoned (correctly) that humans and apes probably are closely related, sharing a common ancestor. Darwin wrote these prescient remarks prior to the major discoveries of early hominins (hominin refers to the genus *Homo* and their upright ancestors, such as *Australopithecus afarensis*). Another group of relatives, the hominids, are the great apes. This group includes humans, orang-utans, gorillas, chimpanzees, bonobos and their common ancestor. The cast in the story of human evolution includes the offshoots of *Homo erectus*, up to modern humans. To understand the relationships between some of these different species and whether or not they spoke, one must learn what is known about them.

Part of the controversy of human origins is the number of species of *Homo* that existed, but it is still necessary to understand the potential cognitive abilities of all hominins (based on brain size, tool kits and travel) before moving on to the significance of hominin migration for the evolution of human language. One can focus on physiology or culture or both, yet some of the most interesting evidence comes from culture.

Symbols (the association of largely arbitrary forms with specific meanings, such as using the sounds in the word 'dog' to mean *canine*) were the invention that put humans on the road to language. And for this reason we must understand not only how they came about, but also how they were adopted by entire communities and how they were organised. One proposal I discard is arguably the most influential explanation of the origin of human language of all time. This is the idea that language resulted from a single genetic mutation some 50–100,000 years ago. This mutation supposedly enabled *Homo sapiens* to build complex sentences. This is the set of ideas known as *universal grammar*. But a very different hypothesis emerges from a careful examination of the evidence for the biological and cultural evolution of our species, namely the *sign progression* theory of language origin. This phrase means simply that language emerges gradually from indexes (items that represent things they are physically connected to, such as a footprint to an animal) to icons (things that physically resemble the things they are used to represent, such as a portrait for the real person) and finally by creating symbols (conventional ways of representing meaning that are largely arbitrary).

Eventually, these symbols are combined with others to produce grammar, building complex symbols out of simple ones. This sign progression eventually reaches a point in language's evolution in which gestures and intonation are integrated with grammar and meaning to form a full human language. This integration transmits and highlights the information that the speaker is telling the hearer about. It represents a crucial, though often ignored, step in the origin of language.

Because the evolution of language is such a hard problem, the earliest efforts to solve it predictably began rather badly. In place of data and knowledge, accounts relied on speculation. One popular idea was that all languages began with Hebrew, since it was believed that this was the language of God. Like this Hebrew-first speculation, many ideas were abandoned, although there were others that included kernels of good ideas. These have led, however circuitously, to the present understanding of language origins.

But a serious deficiency traced its way through all of these early efforts and a lack of evidence, in conjunction with an abundance of speculation, irritated many scientists. So in 1866 the Paris Linguistics Society declared that it would no longer accept papers about language origins.

The good news is that the ban has now been lifted. Contemporary work is somewhat less speculative and occasionally more firmly grounded in hard evidence than the work of the nineteenth and twentieth centuries. In the twenty-first century, in spite of the difficulties, scientists have finally managed to put together enough of the extremely small pieces of the language evolution puzzle to give a reasonable idea of how human languages came about.

Still, one of the greatest mysteries left to solve regarding the origin of language, as many have observed, is the 'language gap'. There is a wide and deep linguistic chasm between humans and all other species. Communication systems of the animal kingdom are unlike human language. Only human languages have symbols and only human languages are significantly compositional, breaking down utterances into smaller meaningful parts, such as stories into paragraphs, paragraphs into sentences, sentences into phrases and phrases into words. Each smaller unit contributes to the meaning of the larger unit of which it is a part. For some, this language gap exists simply because humans are a special creature unlike any other. Others claim that the distinctiveness of human language was designed by God.

More likely, the gap was formed by baby steps, by homeopathic changes spurred by culture. Yes, human languages are dramatically different from the communication systems of other animals, but the cognitive and cultural steps to get beyond the 'language threshold' were smaller than many seem to think. The evidence shows that there was no 'sudden leap' to the uniquely human features of language, but that our predecessor species in the genus *Homo* and earlier, perhaps among the australopithecines, slowly but surely progressed until humans achieved language. This slow march taken by early hominins resulted eventually in a yawning evolutionary chasm between human language and other animal communication. Eventually, *Homo* species developed social complexity, culture and physiological and neurological advantages over all other creatures.

Human language thus begins humbly, as a communication system among early hominids not unlike the communication systems of many other animals, but more effective than a rattlesnake's.

What if all eighty remaining speakers of Banawá died out suddenly and their bones were discovered only 100,000 years hence? Forgetting for now the fact that linguists have published grammars, dictionaries and other studies of the Banawá language, would their material culture leave any evidence that they were capable of language and symbolic reasoning? Arguably it would leave even less evidence of language than has been found for *neanderthalensis* or *erectus*. Banawá art (such as necklaces, basket designs and carvings) and their tools (including bows, arrows, blowguns, darts, poison and baskets) are biodegradable. So their material culture would disappear without a trace in much less than the 800,000 to 1,500,000 years that have passed since the appearance of the earliest cultures. Of course, it might be determined from soil usage that they had villages of a certain size, huts and so on, but it would be as difficult to extrapolate from the remnants of their artefacts that they had language, as it is to claim that many ancient hunter-gatherer groups did or did not have language. It is known that current populations of Amazonians have fully developed human languages and rich cultures, so care must be taken not to conclude prematurely that the absence of evidence about language or culture in the prehistoric record indicates that ancient human populations lacked these essential cognitive attributes. In fact, when we look closely, there is evidence that the earliest species of *Homo* did in fact have culture and did speak.

The solution to the mystery of human language origins begins with an examination of the nature and evolution of the only surviving linguistic species, *Homo sapiens*, or, as author Tom Wolfe puts it, *Homo loquax*: 'speaking man'. There are several unique perspectives that mark the path of the evolution of language.

First, human language emerges from the much larger phenomenon of animal communication. Communication is nothing more than the (usually intentional) transference of information from one entity to another, whether this be the pheromonal communication of ants to other ants, the calls of vervets, the tail positions and movements of dogs, the fables of Aesop, or the writing and reading of books. Language is much more than information transfer, though.

The second perspective on the evolution of language derives from an examination from both the biological and cultural vantage points. How did the brain, the vocal apparatus, movements of the hands and the rest of the human body, in conjunction with culture, affect and facilitate language evolution? Too many accounts of language evolution focus on one or the other of these, the biological vs the cultural, to the exclusion of the others.

A final, and necessary, perspective may strike some as curious. It is to look at language evolution as a linguistic field researcher would. That perspective leads to two fundamental questions: how similar are the human languages that are spoken today and what does the diversity of modern languages reveal about the first human languages? These perspectives offer a useful vision of evolutionary milestones that mark the path of the first language of *Homo* species.

There are still additional questions to answer. Are gestures crucial to human languages? Yes, they are. Is a vocal apparatus identical to that of modern humans necessary for human languages? No. Are complex grammatical structures required for human languages? No, but they are found in many modern languages for a variety of reasons. Do some societies communicate less or use linguistic communication less than others? It seems so. *Erectus* might have been in possession of language yet nevertheless valued taciturnity.

Part One

The First Hominins

1

Rise of the Hominins

The hand of the Lord was on me, and he brought me out by the Spirit of the Lord and set me in the middle of a valley; it was full of bones. He led me back and forth among them, and I saw a great many bones on the floor of the valley, bones that were very dry ...

Ezekiel 37:1–2

CONTROVERSY IS OFTEN DIFFICULT to resolve. In June of 2011 a young mother, Casey Anthony, was on trial for the murder of her two-year-old daughter, Caylee Anthony. The prosecution supported its allegation that Casey murdered her daughter with evidence that her daughter's body had been stored in the trunk of Casey's car – a car only she had access to – for several days in 90-degree weather. They produced witnesses who claimed that they had smelled the stench of a decomposing body in the trunk of that car and also showed that there were bugs in the trunk typical of those that would have swarmed and multiplied in the hot sun on a dead body. Grisly evidence, to be sure. But it sounds convincing. Had the trial stopped there, perhaps a guilty verdict would have been rendered.

First, however, the defence needed to plead their case. Of course, they called their own witnesses, including a forensic expert who argued that the smell people reported could have come instead from a bag of garbage that Casey had left in her trunk for more than a week (no one was defending her hygiene). Moreover, the forensic witness claimed that the bugs found in the trunk of Casey's car were neither of the type nor in the quantity that would be expected if her car trunk had contained a decomposing body. Finally, after much more arguing back and forth between the experts and attorneys, the jury ultimately decided in favour of the defence. Twelve people found the defendant's story sufficiently credible to raise reasonable doubt about what happened to Caylee.

The problem of reasonable doubt that faces some juries is also common in science. But the difference is that scientists, unlike jurists, *thrive* on reasonable doubt. This is because, like doubt, they are trying neither to convict theories, nor to exonerate them. Rather, scientists want to *evaluate* theories, rejecting those that have *excess* reasonable doubt, even if only temporarily. In other words, doubt is an intellectual tool that allows scientists to narrow down the number of theories they need to concern themselves with.

It is unsurprising that disagreement between experts occurs. In fact, consensus among experts often seems rarer than disagreement. Every scientific advance usually originates as a dispute concerning the interpretation of evidence for vs evidence against some thesis. Science is not about finding a 'true' theory. It is about finding the best theory, as scientists grope their way towards understanding.

Much vaster and more complicated than any murder trial is the quest to understand the origin of humans and their languages. This effort requires a picture of the trajectory from the initial state of hominins to the current state, and is always going to be fraught with controversy and disagreement. Definitive knowledge is lacking even on such basic questions as the variability in the complexity of human reasoning across human species in the evolutionary record. There is not even consensus on the range of variation in the five 's's: smarts, speed, size, sex and strength among modern humans.

So why are such problems regarding the limits of human capacity relevant to understanding the species' evolution? Because specialists and laypeople alike fail to agree what new evidence means since they *interpret* any new discoveries or findings differently. Rather than naively anticipating agreement, one can hope instead for a weighing of alternative accounts. Most specialists are able to determine when one account has cast reasonable doubt upon another. But no one can tell someone which account to choose, nor can they predict which account someone will select. Scientific choices are intellectually, culturally and psychologically motivated.

Part of understanding human species surely must be to appreciate how humans came to achieve greater cognitive success than any other species. Humans are everywhere. Like cockroaches and rats, they are adaptable, multiply quickly and travel well. They are tough and resilient. They are clever. They can be territorial, diurnal, nocturnal, or crepuscular. They can be kind or vicious. Humans have become, for better or worse, lords of the planet. If the dinosaurs were still alive today, humans would kill them for trophies, or eat them, or put them in parks and zoos. They would be no match for *sapiens*. Humans, not they, are the apex predators of all time on this planet. This success has much to do with the fact that, though *sapiens* are small with soft skin, no claws or serious strength, they talk to each other. Because humans can talk they can plan, they can share knowledge, they can even leave knowledge for future generations. And therein lies the human advantage over all other terrestrial species.

So exactly what is this ability of humans – what is language? It isn't possible to talk about how some characteristic, such as language, evolved without at least some idea about what this characteristic is.

Language is the interaction of meaning (semantics), conditions on usage (pragmatics), the physical properties of its inventory of sounds (phonetics), a grammar (syntax, or sentence structure), phonology (sound structure), morphology (word structure), discourse conversational organisational principles, information and gestures. Language is a gestalt – the whole is greater than the sum of its parts. That is to say, the whole is not understood merely by examining its individual components.

Indeed, there are entire communities of linguists who identify themselves by the different subareas. There are pragmaticians, conversational analysts, syntacticians, morphologists, phoneticians, semanticists and so on. But none of them is studying language as a whole, only the parts they are interested in professionally. A syntactician is to language as an ophthalmologist is to the body. Both are necessary, but each is (understandably) tackling a very small piece of the pie.

What is the full pie supposed to be like, then? It is a communication system. And this is what the evolutionary and contemporary evidence points to – namely that the ultimate purpose and accomplishment of language is the building of communities, cultures and societies. These are built through stories and conversations, written or oral, each of which helps to establish and justify shared value priorities for cultures or individuals. Language, in fact, builds the knowledge structures that are peculiar to a particular culture (such as the colours recognised, the types of professions considered most attractive, medical understanding, mathematics and all of the other things humans know as members of a society). And language also helps to interpret the different social roles, such as father, boss, employee, doctor, teacher and student, that a culture recognises.

Grammar is a tremendous aid to language and also helps in thinking. But it really is at best only a small part of any language, and its importance varies from one language to another. There are tongues that have very little grammar and others in which it is extremely complex.

The course followed by humans on the path to language was a progression through natural signs to human symbols. Signs and symbols are explained in reference to a theory of 'semiotics' in the writings of Charles Sanders Peirce. C. S. Peirce was perhaps the most brilliant American philosopher in history. He contributed to mathematics, to science, to the study of language and to philosophy. He is the founder of two separate fields of study: *semiotics* – the study of signs – and *pragmatism* – the only uniquely American school of philosophy. In spite of his brilliance, he was never able to secure long-term employment because he was cantankerous and rebellious against social mores. Peirce's semiotics did not concern itself directly with the evolution of language. But it turns out to be the best model of the stages of linguistic evolution.

Peirce's theory indirectly predicts a progression of signs from natural signs (indexes), to icons, to human-created symbols.* This progression moves to an increasing complexity of types of sign and the evolutionary progression of *Homo* species' language abilities. A sign is any pairing of a form (such as a word, a smell, a sound, a street sign, or Morse code) with a meaning (what the sign refers to). An index, as the most primitive part of the progression, is a form that manifests an actual physical link to what it refers to. The footprint of a cat is an index: it indicates, makes us expect to see, a cat. The smell of a grilling steak brings the steak and the grill to mind. Smoke indexes fire. An icon is something

---

*I deviate slightly from Peirce here. For Peirce, indexes were more complex than icons, as used and elaborated by humans. But as used by non-humans and in evolution, I believe that indexes precede icons.

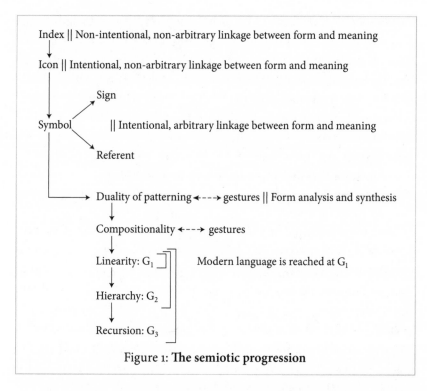

Figure 1: **The semiotic progression**

that physically evokes what it refers to: a sculpture or portrait references its subject via a physical resemblance. An onomatopoeic word like 'bam' or 'clang' brings to mind those sounds.

Symbols are conventional links to what they refer to. They are more complicated than other signs because they need not bear any resemblance to nor any physical connection to what they refer to. They are agreed upon by society. The numeral '3' refers to the cardinality of three objects just as 'Dan' refers to someone of that name, not because the word 'three' bears a physical connection or resemblance to cardinality, nor because all people named Dan have any physical characteristic in common. This arbitrary, conventional association of form and meaning is exactly what renders symbols the beginning of language and evidence for social norms. Symbols are the original social contract.

Figure 1 gives us a view of the relative order of language evolution, following Peirce's ideas closely (though putting the index before the icon).

Once humans had symbols, certain portions of those symbols became more meaningful than other parts. If I randomly choose to write an S as Ŝ or a P as $\mathcal{P}$ in English prose, native English readers will disregard the embellishments, recognising immediately that the additions are irrelevant. But if I write S as P, then this will cause confusion. This is because the meaningful parts of symbols cannot be changed without obscuring their identity, though non-meaningful parts may change with impunity. Symbols are, therefore, not simplex atoms but contain 'junk' portions (the parts that are non-essential to their meaning) alongside crucial information. For Peirce this information was crucial for the 'interpretant' or meaning of the sign.

From symbols and interpretants, it is a short step for language to progress to the phenomenon known as 'duality of patterning', which organises smaller units into larger items. Duality of patterning enables the transition to the three levels of complexity – G_1, G_2 and G_3 – that distinguish the different types of grammars that human societies may choose to build their languages around, shown in Figure 1.

The hypothesis that language evolution followed the increasing level of complexity in Peirce's signs is supported by the archaeological record. On the other hand, the jump from icon to symbol in this chart is 'unnatural'. This step requires human invention. Evolution did not create symbols or grammars. Human creativity and intelligence did. And that is why the story of how language began must also be about invention rather than about evolution alone. Evolution made our brains. And humans took over from there.

Still, it is necessary to discuss much more than language itself to understand what happened. For that, one must link language's development to the species' biological development. According to the available evidence, *Homo sapiens*, like every family, fits into a particular set of relationships, usually referred to as a phylogenetic tree or clade (Figure 2). This is highly speculative, but it does give us a reference point. The lower branches among *Homo* species may turn out to be quite different than those given here.

All animals communicate, hence the arrow at Animal Kingdom. Not all animals have language, which only seems to emerge through the evolution of the genus *Homo*.

Like anything complicated and controversial, there are a number of ideas about how life began on earth.* One prevalent proposal is

---

*This chart provides more detail than we will refer to elsewhere in this book. It

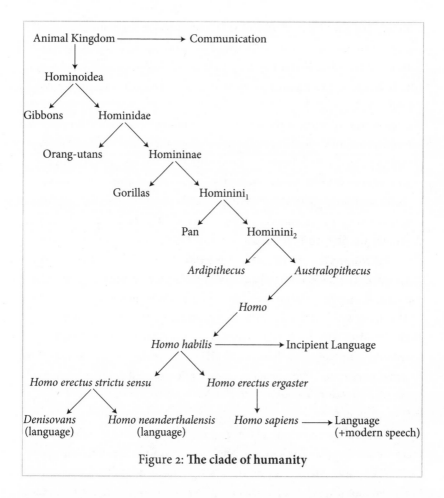

Figure 2: **The clade of humanity**

that a supreme deity created life. Any discussion of a theistic account of life and language must acknowledge that the origin of both is an important line in the sand for theists. One frequent theistic answer to the question of how DNA and subsequent forms of life evolved is the 'watchmaker' theory. Watches were, at the time of this metaphor, the highest technology known. For many reasons, discussions of philosophers and theists often revolve around the most advanced technology of the day. In this case, watches are intricate, complicated, hierarchical in structure and obviously designed. So if someone found a watch on

should be seen as implicit in the primate chart of Figure 3.

a distant planet, the presence of that watch would indicate that somewhere there was a designer who had a purpose in mind for it, designed
it and fabricated it. William Paley put it this way in his *Natural Theol-
ogy* in 1802:

> In crossing a heath, suppose I pitched my foot against a stone, and
> were asked how the stone came to be there; I might possibly answer,
> that, for anything I knew to the contrary, it had lain there forever:
> nor would it perhaps be very easy to show the absurdity of this
> answer. But suppose I had found a watch upon the ground, and it
> should be inquired how the watch happened to be in that place; I
> should hardly think of the answer I had before given, that for any-
> thing I knew, the watch might have always been there ... There must
> have existed, at some time, and at some place or other, an artifi-
> cer or artificers, who formed [the watch] for the purpose which we
> find it actually to answer; who comprehended its construction, and
> designed its use ... Every indication of contrivance, every manifes-
> tation of design, which existed in the watch, exists in the works of
> nature; with the difference, on the side of nature, of being greater or
> more, and that in a degree which exceeds all computation.

Paley's argument for an 'artificer' precedes the work of Wallace and
Darwin on evolution by natural selection by more than half a century.
There are modern theologians and theistic scientists who consider this
argument sound, substituting a complex organ such as the eye in place
of the watch. But philosopher David Hume pointed out three serious
problems with the watch analogy. First, the materials of the watch are
not found naturally – the watch is built from human-made materials.
This makes the analogy artificial. As Hume said, it would make much
more sense to use something composed exclusively of organic materials, such as a squash, instead of a watch because one can observe that
squashes come forth on their own.

Hume's second objection is that one may not use experiential knowledge to infer a conclusion about non-experiential knowledge. If you
understand what a watch is, you also know that the watch was created.
One could even observe a watch being made. Yet no one could have any
direct experience with the creation of the world. Thus the conclusion

that because a watch has a designer the universe also does is not only empirically unjustified but also illogical. Finally, Hume remarked that even if a watch did show that every complicated thing, the universe in particular, has a designer, this lesson would still have nothing to say about the nature of that designer. Such reasoning thus, even if it had not been shown to be invalid, supports no known religion or idea of a deity above any other.

Perhaps the most effective argument against the watchmaker analogy, however, comes from culture. No person can make a watch or its component materials by themselves. A watch is the output of a culture, not a designer. If the universe was designed, this design would have required a society, not a god, unless that god were far different than it is described in the major religions. More to the point, however, the theistic design argument for the universe fails because the science says so. There is a solid scientific foundation to evolutionary theory that is lacking in theistic accounts.

Evolution is a well-established *fact*. Only the explanations of how evolution happens or looks – natural selection, genetic processes and family trees – can be called theories. But evolution itself is not a theory. In order to understand the origin of language, the origin of life more generally must be considered in order to frame the discussion. And that requires evolution.

The earth is roughly 4.5 billion years old and probably began as a whirling cloud, cooling and solidifying, its waters gradually ending their cycle of rain and evaporation, reducing the surface temperature of our red planet enough to accumulate in turbulent, lifeless, hot oceans inhospitable to all life, in the Precambrian period's Archaean Eon. But phosphates, sugars, nitrogen were about in the oceanic stew, and from these the first carbohydrates and other building blocks of life were formed. At least, that is one possible explanation.

Another account offered by scientists is that replicating DNA wholly originated in space and was brought to earth by a meteorite or asteroid. This proposal for the origin of DNA is known as 'panspermia'. According to the proponents of this view, nucleotides are more easily formed in the cold and ice of comets. Our planet was like a giant ovum floating in space, fertilised by space dust, meteors and asteroids, the spermatozoa of the universe, which brought DNA to us. There is even some convincing evidence for this view. Meteorites regularly enter earth's atmosphere. Some of these might have brought DNA to earth from another part of the universe or solar system. Or perhaps meteors brought not DNA but nucleotides from other parts of space.

Whatever happened, nucleotides eventually joined together in the seas. Later, membranes began to form around them. Within those membranes nucleic acids, along with ribonucleic acid (RNA) and deoxyribonucleic acid (DNA) formed. At some point these acids took on the property responsible for all life – replication. From the formation of the earth until molecular life began to form took roughly 500 million years. Within another 500–800 million years, life forms large enough to be visible as fossils were produced.*

From the beginning of this nucleotide soup, earth transitioned to the Proterozoic 'early life' period of the Precambrian. The foundation of life, DNA, again formed from sugars, phosphates and nitrogen.

Because of an understanding of DNA, it is known that a human and a dog are distinguished at the molecular level by the composition of their DNA and the ways in which their DNA is sequenced to form their genome. Genomes are thus the sum of the various DNA and RNA and their combinations. Fine-tuning a bit, canines and humans are not merely distinguished by the components of their DNA but by the *syntax* of their DNA. The hierarchy of DNA and units relevant to it is:

Chromosomes (carriers of DNA)

↓

DNA + histones†

↓

Genes (segments of DNA)

---

*There are other hypotheses on the origins of life. One that is widely supported, though by no means universally accepted, is the so-called 'RNA world hypothesis'. According to this hypothesis, since the essence of life is self-replication and since RNA has this property, the presence of RNA would have preceded both proteins and DNA. DNA would have come later to provide storage or memory capacity to the RNA and proteins would have eventually been synthesised, taking over some of RNA's functions, though of course RNA continues to be essential to life as we know it.

†Histones are the packaging around DNA that controls how genes are activated and deactivated.

If one considers the earth from its beginning until today, 99.997 per cent of the planet's history had passed before the Pleistocene (2.8 million years ago), when *Homo habilis* (or *Homo erectus*, depending on classification) the first species of the genus *Homo*, emerged. Species such as *sapiens* arose even later. The period in which the earliest hominins gave rise to modern humans, began from the late Miocene (23–5.3 million years ago), through the Pliocene (5.3–2.8 million years ago) and the Pleistocene, up until the current age, the Holocene (11,000 years ago).

Humans enjoy a privilege unique among all other life forms that enables them to contemplate their origin. And yet, all human perspectives are culturally shaped. Therefore, along with superior mental powers, cultures not only guide humans' understanding of the world but also define what is worth looking at. Culture constrains how humans justify their reasoning. Science emerges from and is shaped by the values of culture, different social roles and the knowledge structures that have been sanctioned by society (that is, what knowledge is and how different components of our knowledge are related to one another).

Culture is one reason that different scientists take different views on the fossil evidence. It is not merely disagreement about the facts, though that too is important. Richard Feynman was one of the first to notice that results of physics experiments tended to be closer to published expectations than one would have otherwise expected. This points to one cultural effect in science known as 'confirmation bias'. Even though consideration is given only to science, there is no escaping from the shadow of cultural influences. Interpretations of much of the fossil record change regularly. My conclusions here are no different, although this difference makes them no better or worse than other conclusions until more data is brought to bear.

The accumulated scientific record built by *Homo sapiens*, through language and Western culture, concludes that humans are primates and that the roots of their genus are to be found in the origins of the primates. So, what are primates? And where did they come from?

The primate order, of which *Homo sapiens* is one species, arose more than 56 million years ago. Because evolution is gradual and continuous there are 'protoprimates' that precede the 'proper primates'. The earliest known protoprimate transitional fossil is *Plesiadapis tricuspidens*,

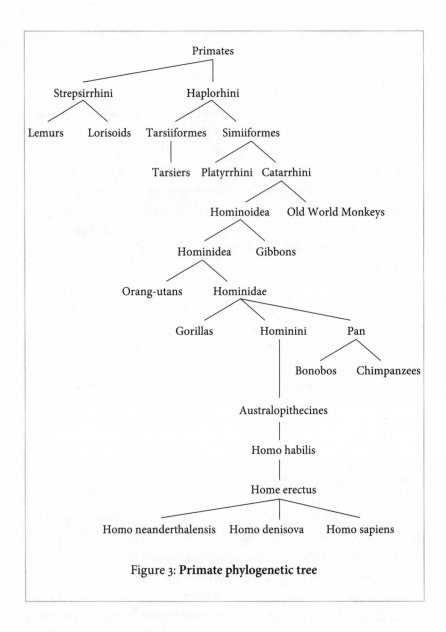

Figure 3: **Primate phylogenetic tree**

which existed some 58 million years ago in North America.

But the story of primate evolution begins in earnest with the first-known true primate genus, the *Teilhardina*, which were the precursors of all other primates, including us. The *Teilhardina* (named after the Jesuit theologian and palaeontologist Pierre Teilhard de Chardin) were small creatures, similar in size to modern marmosets. As these creatures developed and found separate niches, they formed varieties and eventually new species. From the new species came new genera.*

Many primates break down into genera according to the shape or conditions of their noses. For example, the wet-nosed or strepsirrhine primates today are limited almost exclusively to the lemurs of Madagascar. Humans are counted among the haplorhini or 'dry-nosed' monkeys. The haplorhini in turn break down into the tarsiiformes and simiiformes – simians, or anthropoids – from which humans come. There is further subdivision into catarrhines (long noses) and platyrrhines (flat or 'downward' noses). All monkeys native to the New World have nostrils that point sidewards, while all Old World primates, including humans, have nostrils that point downwards. Figure 3 summarises the classification of humans among other primates.

Figure 3 shows the primate tree with humans as part of the infraorder catarrhini, as previously described. As well as being catarrhines, humans are also, along with all monkeys and apes, simians. The platyrrhines of the Americas evolved differently from their Old World relatives, producing no apes, only monkeys. No humans therefore evolved in the New World.

All primates can climb trees. Though I confess that I cannot climb as well as a woolly monkey, all humans possess, as primates, bodies designed at least partially for tree climbing. Some also stand out from other mammals for more important reasons than their climbing prowess: humans have large brains, a reduced sense of smell and stereoscopic – 3D – vision, all traits that have helped them become the rulers of a very real planet of the apes.

Evolution is the great procrustean bed of life, where species are stretched, chopped and otherwise modified to fit their niche. As evolution brought the primates forth from the 'great mammalian surge' that followed the end of the dinosaurs during the Cretaceous–Paleogene (K–Pg) extinction event some 66 million years ago, it launched another in a long line of cognitive leaps. Eventually, this long mammalian advance produced the hominids.

Genera is the plural of *genus*, which is a set of species sharing an immediate common ancestor.

Most modern scientists use the term hominid, the singular form of the Latin word *hominidae*, men, to refer to all the great apes (the English form of the plural is hominids). There are a few who still prefer to use the word in its older sense, to refer not to all the great apes, but to the genus *Homo*. But most modern scholars reserve hominid to describe collectively the great apes and hominin (hominini), humans and all of their direct ancestors, beginning from the time that chimps and humans split into separate branches of the primate tree. Using two terms, hominid and *Homo*, does not increase the perceived complexity of human evolution; it simply helps to avoid ambiguity.

By contrast, Darwin's theory of evolution by natural selection is much less complicated. Indeed, part of its attraction and elegance is its very simplicity. It consists of three postulates:

First, the ability of a population to expand is infinite in principle, though the ability of an environment to support a population is always finite.

Second, there is variation within every population. Therefore, no two organisms are exactly alike. Variation affects the ability of individuals to survive and multiply as some produce more viable offspring in a given niche.

Third, somehow parents are able to pass their variation along to their offspring.*

Together, these three postulates have come to be combined as 'descent by natural selection'. Darwin noticed during his famous trip aboard the HMS *Beagle* (1831–1836) that animals seemed adapted to fit their environment. And throughout his scientific career he also noticed how different creatures (such as his famous Galapagos finches) adapt rapidly to even small changes in their environments.

Ironically, while Darwin was proposing natural selection, a Czech monk, Gregor Mendel, was laying the foundations of genetic theory. Working with garden peas for seven years (1856–1863), Mendel developed two important principles of genetic research:

First was the principle of 'segregation', which is the idea that given

---

*Unfortunately Darwin's ideas about how traits are 'passed along' differ greatly from what we now know to be true about genetics, which is hardly surprising given that he died before modern genetics had been developed.

traits are broken into two parts (known now as alleles) and only one of these parts passes from each parent to their shared offspring. Which allele is passed along is random.*

Second came the principle of 'independent assortment'. By this principle the pairs of alleles produced by the union of the parents' haploid cells form new combinations of genes, that are not present in either the mother or the father.

Though Mendel's work eventually gained dramatic acceptance, there were still problems that were overcome only with the advent of modern genetics. First, as Thomas Morgan showed in his now famous work on fruit-fly mutations, genes are linked; that is they work in tandem in many traits. This is contrary to Mendel's idea that each gene behaves independently, an idea that many geneticists had embraced prior to Morgan's work. Morgan's findings meant that the independent assortment of genes wasn't right – *contra* Mendel. Morgan also brought to his research original and long-term interests in cell structures. Through this work it became clear that chromosomes were real entities in cells and not merely hypothetical gene vehicles.[1] Second, Mendel suggested that variation is always discrete, but in fact it is usually continuous. If one mates a 6'5" mother with a 5'2" father one will not get simply either a 5'2" vs a 6'5" child, but any height between those at minimum, among other possibilities. In other words, many (in fact most) traits blend. This is not captured in the simple interpretation of independent assortment from Mendel's work on peas, in which all traits he worked on were discrete and, for the purposes of his research, unrelated.

Another crucial fact about evolution is that the targets of natural selection are phenotypes (externally visible physical and behavioural attributes, resulting from genes and environment), not genotypes (the genetic information that is partially responsible for the phenotype). Thus natural selection operates on (selects for survival) creatures based on their behaviour and overall physical properties. Genes underlie

---

*We now know something that Mendel did not, namely that individual alleles are selected by the process of *meiosis*, in which a *haploid* cell (a cell with only half of the normal number of chromosomes for a particular species) is formed. Two haploid cells (ovum and sperm) are contributed, one by each of the two parents, before reproduction begins.

these properties and behaviours, but the phenotype is more than genes: it is partially produced by histones, environment and culture. The histones control the timing of the unfolding of genetic information and thus how the genes produce the phenotype.

When non-biologists think of evolution, they often conjure up ideas of new species, but, though that is one by-product of evolution, looking only at speciation as evidence is misleading. If a creation 'scientist' says that creatures can't transform into others, and that evolution must therefore be false, what they are actually disputing is *macroevolution* – evolution on a grand scale. However, macroevolution is not the only form of evolution. In fact, macroevolution is usually the accumulation of smaller evolutionary changes, perhaps as small as the mutation of a single allele, known as *microevolution*.

While microevolution is, by definition, less discernible, especially to observers with short human lifespans, it is where the real action takes place. As a result, if one can explain the small changes, the larger changes will follow by and large. Evolutionary scientists seek to understand biological change over time (evolution) in *all* of its forms. Macro- and microevolution are simply points along a vast continuum of modification by natural selection.

One of the ways in which micro- and macroevolution are stimulated is via mutation. Many mutations are neutral. Other mutations are fatal. But some mutations provide a survival advantage to their host organism. A change favoured by natural selection in a particular environment is advantageous if the mutated creature produces more viable offspring than creatures lacking the mutation.

Neutral mutations are important for evolutionary theory even though they are by definition neither harmful nor helpful to the survival of their host. As Linus Pauling, the only person in history to win two unshared Nobel prizes, one for chemistry and one for peace, and Emile Zuckerkandl, a pioneer in genetic dating, proposed in 1962, neutral changes occur at a constant rate over time. This constancy works like a molecular clock, one that can help determine when two related species diverged. Today it has become a vital tool in understanding evolutionary differences among creatures, even though they are not themselves responsible for those differences.

Mutations favoured by natural selection, however, are not the only way that evolution works. For anything as complicated as the whole of life on earth, it should not be surprising to learn that no one concept explains it all. There are other sources of micro- and macroevolution than natural selection. One of these is known as 'genetic drift'. Technically, in genetic drift there is a reduction in a population's genetic diversity. Imagine that the population of all humans is one thousand individuals from one hundred families. Now assume that the genes that produce photopigments in five families from this one thousand, say fifty individuals, are deficient. These individuals are 'colour-blind'. Next, imagine that these colour-blind individuals come to be shunned by the majority as undesirable for some cultural reason and that all fifty of them therefore decide to move elsewhere. Finally, suppose that the original population, the non-colour-blind individuals, are wiped out by disease or natural disaster after the departure of the colour-blind individuals. The colour-blind people are unaffected. This improbable but possible chain of events will result in a state where the only genes left among the species are the genes that produce colour-blindness. The colour-blind community may grow over time, producing many off-spring and descendants, founding entirely new populations of humans. This scenario would result in significant changes to the human species, independent of natural selection.

Genetic drift is a naturally occurring reduction in genetic diversity that is produced by Mendel's principle of randomness in the selection of alleles. Again, this is not caused by natural selection because fitness plays no role in the result.

A special case of genetic drift is known as a population bottleneck. A population bottleneck is an alteration in the allele ratio produced by external causes, as in our example of ostracised colour-blindness sufferers. Such a bottleneck can include things like migration, where a migrating population is some sample of the main population wherein there is a different allele ratio than that found in the population as a whole. Population bottlenecks include any reduction of the genetic diversity of a population caused by external events. Take a disease that kills off one member of each family. Chances are that reduction left a different gene distribution in the overall population, thus producing a population bottleneck. This too can lead to a 'founder effect' – a subpopulation with a different distribution of alleles than the original population that in turn produces generations of viable offspring. In other words, if the original population of *Homo erectus* that left Africa had a different allele ratio than the population that remained in Africa, the former and latter populations would be separate founder populations for the ensuing generations.

Another form of evolutionary change is that effected by culture, a form known as the 'Baldwin effect', and particularly relevant to the evolution of human language. The Baldwin effect, first proposed in 1896 by psychologist James Mark Baldwin, was an important conceptual advance in evolutionary theory for at least two reasons. First, it underscored the importance of phenotypes (visible behaviours and physical characteristics) for natural selection. Second, it demonstrated the possible interaction of culture with natural selection. As a hypothetical example, let's suppose that a population of *Homo erectus* enters Siberia in the summer only to discover later that Siberia is cold in the winter. Now assume that everyone learns how to make winter garments from bear fur and that the most effective stitching of these furs requires manual dexterity that is extremely challenging or unavailable to the community as a whole – except for one lucky person who has a genetic mutation that allows him or her to bend their thumb to their forefinger in such a way as to produce a more effective, long-lasting stitch for bear fur coats. They therefore make more effective coats for their family. This in turn allows the members of their family to produce more offspring than the families of the stitch-challenged. Eventually, the mutation will increase the chance that the original mutant's 'dexterity genetics' will reproduce through their offspring who, in turn, out-survive (in winter at least) the offspring of the less dexterous stitchers. Over time, the dexterity gene will spread throughout the population.

The same genetic mutation in another environment would not have propagated throughout the population because it might not have provided any survival benefit. It would be in another environment simply a neutral mutation, of which there are many. This might happen if the coat-making phenotype is neutral in a warmer climate, such as Africa. We can say, therefore, that culture can turn neutral mutations into positive mutations. The Baldwin effect, also known as dual inheritance theory, brings culture and biology together and seeks to explain those evolutionary changes that can't be explained by either one on their own.

Now using our imagination once again, let's suppose that a woman is born during the time that humans are developing language. We will call this woman Ms Syntax. While the rest of the community says things like, 'You friend. He friend. She not friend,' Ms Syntax says, 'You friend and he friend but she not friend.' Or while everyone else is talking like, 'Man hit me. Man bad,' Syntax Lady might say, 'The man who hit me bad.' In other words, the syntax master has the ability to make complex sentences while the rest of the population can only form simple sentences. Could the entry of complex sentences into human language have been a mutation, spreading via the Baldwin effect or some other mechanism, such as sexual selection? This is unlikely. Language presents a different case than genes for physical skills.

The first reason for doubting that a mutation for syntax could spread through a population or be favoured by the Baldwin effect is that it is unlikely that complex sentences would provide a survival advantage, especially in light of the fact that there are languages spoken today, as we discuss later on, that lack such complex syntax. These latter languages have survived in the same world as languages that do have complex syntax. Moreover, even if it were discovered that the languages currently claimed to lack complex syntax did show such syntax in some cases, this discovery would only underscore the fact that speakers of these languages survive fine in an environment 99 per cent free of complex sentences.

More importantly for Ms Syntax, in order to be able to *interpret* complex sentences, one would need to be able to interpret complex syntax. Uttering complex sentences in a population that lacks the ability to do this – where, in other words, you are the only person that is able to interpret or produce complex utterances – would be like yelling a warning to a deaf, mute and blind person. One could argue that non-human primates can already do this, since they seem to be able to respond effectively to requests using complex syntax (the bonobo Kanzi comes to mind). But this is a far cry from actually being able to fully understand complex sentences. Following instructions given in recursive sentences, for example, might be a first step towards acquiring or evolving recursion, but only that. Ability to think in complex ways must precede talking in complex syntactic constructions or no one would be able to fully understand those utterances.

But how might such thinking arise? How could someone think in ways that they do not speak? One possibility is, perhaps, by planning events within events via images within images or even, as many speakers in the world seem to do today, by thinking in larger stories that, although they use simple sentences, weave together complex thoughts:

John fishes.
Bill fishes.
John catches fish.
Bill stops.
Bill eats John fish.
Bill returns.
John returns same time.

This story, completely composed of non-complex sentences, says that John went to fish and later, or at the same time – depending on which is inferred by the context – Bill went to fish. John caught fish before Bill. So Bill stopped fishing and ate fish with John. Bill decided to stop fishing and return home. John returned home with him. There are, in fact, many languages in which simple sentences are woven together in complex stories just like this.

Another example of complex thinking without complex sentences is intricate task performance or planning without talking at all, such as in weaving a basket with many parts. As the hypothetical fishing story has just shown, complex sentences are not required for complex thinking or storytelling. Complex thinking might make it possible to utter and interpret complex sentences, but it doesn't require them. The reverse, however, is not true. One must be able to compose complex meanings in order to interpret a complex sentence.

On the other hand, it is possible that complex syntax would spread through a population by sexual selection. Members of the opposite sex might like to hear the melodic cadences of complex sentences and so might mate more frequently with Ms or Mr Syntax, spreading the syntax genes. But this is unlikely. Complex sentences normally require words that indicate that they are complex. Those words, however, are largely unintelligible outside of the complex syntax they have arisen to signal.

For example, 'John and Bill went to town in order to buy cheese' is a complex sentence, because of its clause-within-a-clause, 'in order to buy cheese', but also because of its coordinate subject noun phrase, 'John and Bill'. The word 'and' is not understandable apart from being able to think in complex syntax. Complex sentences themselves also require complex gestures and pitch patterns that would need to have come about separately and that are unlikely to have arisen owing to a single genetic mutation.

What seems more likely is that complex thinking was favoured by natural selection and thus was a genuine Baldwin effect because it enabled complex planning. It might – and probably would have – shown up later in the form of complex sentences in some languages. In any case, the fascinating conclusion is that natural selection would quite possibly have treated a gene for syntax in language as a neutral mutation, not subject to natural selection.

Recursion, which is a crucial aspect of human thinking and communication, certainly would have arisen early on in human cognition. This is the ability to think thoughts about thoughts, such as, 'Mary is thinking that I am thinking that the baby is going to cry,' or, 'Bill is going to get upset when he finds out that John believes that his wife is being unfaithful to him.' Recursion is also seen in the ability to break tasks down into other tasks, such as, 'First build the spring. Then place the small spring inside a lock. Then place the lock inside another lock and a spring within the larger lock.' And it is visible in complex syntax in sentences: 'John said that Bill said that Peter said that Mary said that Irving said …' It is not clear whether any non-human species is able to use recursive thought, nor whether *erectus* or *neanderthalensis* spoke recursively. But it would not have been necessary for them to have recursion to have language, at least according to the simple idea of language evolution as a sign progression and supported by some modern languages.

It is not difficult to imagine a scenario in which complex or recursive thinking might arise. Suppose that someone is born with the ability to think recursively. This ability to think (not necessarily speak) recursively would provide a cognitive advantage over other members of their community, enabling them to think more strategically, more quickly and more effectively. They might become better hunters, better defenders of the community, or makers of complex tools. This indeed could help them to survive and would in all probability make them more attractive than the competition to the opposite sex, leading to more mating and more offspring. It could also lead to children with the ability to think recursively. And soon this ability would spread throughout the population. At that point only would it be possible to speak recursively and therefore for this new property to be incorporated into the grammar of the community. In other words, there could never be a gene for recursive syntax, because what is needed is a gene for recursive thinking across cognitive tasks. Recursion is a property of thought, not language per se.

Thinking through the implications of these various sources of change in a species, it becomes clear that a single mutation in a particular behaviour such as language would be unable to guarantee that all humans possess the same genetics that they had at the time the original change was introduced. The genotype could have been altered by the Baldwin effect, by genetic drift, or by a population bottleneck.

Still, there is another force in evolutionary theory that can exercise a role in the spread and modification of languages over time. This is 'population genetics'.

Population genetics is concerned with the distribution and frequency of alleles within an entire population. How do groups adapt to their environment? How are new species formed? How are populations divided or structured? Population genetics is one of the most challenging areas of evolutionary theory because of the mathematical sophistication it requires for its application – controlling many variables in many individuals and many links between variables and individuals simultaneously.

One of the pioneers in this field was a postdoctoral associate of Morgan, Theodosius Dobzhansky. Dobzhansky built a bridge between micro- and macroevolution. Basing his studies of populations in their natural habitats, Dobzhansky showed that these populations, however similar they were phenotypically, manifested large degrees of genetic diversity below the surface. Though this genetic diversity is unseen, Dobzhansky demonstrated that is it is always there and is crucial as particular subpopulations differ in their genetic make-up. This diversity renders each subpopulation subject to distinct phenotypic adaptations and speciation, given the right pressures and constraints on gene flow (frequency of mating between the subpopulations).

Dobzhansky was but one of many researchers who examined cross-breeding and genetic drift in small populations and how these could push populations away from one 'adaptive peak' – a kind of local equilibrium in which the environment and the organism match well for a period of time. The basic ideas of population genetics turn out to be crucial to the understanding of change in individual languages and groups of languages over time.*

An overview of the various subfields of genetics and evolutionary change leads to the recognition that fossils are not the only pieces to the puzzle of human origins. The resources of molecular biology are also needed for a full picture. As the genomes of a variety of primates are sequenced, we can begin to estimate dates at which the different lineages of primates diverged evolutionarily. Therefore, because we know that humans and chimpanzees share 96 per cent or so of their DNA sequences, closer than any other two primates, it then follows that there was a common ancestor of humans and chimps unshared by other great apes. Further work leads to the conclusion that this common ancestor split off from other great apes about 7 million years ago. Humans are thus one of the newer apes. From this lineage it is clear that all humans originated, as Darwin predicted, from Africa.

So we know quite a bit about early humans and the story of how life came to be on our planet. But how do we know this? More than DNA evidence is necessary. The reconstruction of the evolution of our species requires the sweat of hard fieldwork, finding, studying and classifying fossils. Here evolutionary theory takes on characteristics of an adventure novel. Who were these fossil hunters? What did they do for our understanding of human evolution? And how did competition and cooperation among them advance the science behind the evolution of human language?

---

*In addition to Dobzhansky's work, that of many others was also important. This work led to what came to be known as the 'new synthesis' in biology and among its leading researchers were Ronald Fisher, especially in his 1930 book, *The Genetic Theory of Natural Selection*, and Sewall Wright, as in his 1932 concept of 'adaptive landscape'.

2

The Fossil Hunters

What we do see depends mainly on what we look for … In the
same field the farmer will notice the crop, the geologists the fossils,
botanists the flowers, artists the coloring, sportsmen the cover for
the game. Though we may all look at the same things, it does not
all follow that we should see them.

John Lubbock

THOUGH DARWIN INITIALLY SEEMED WRONG in his Africa-first
hypothesis, the first evidence that he might be right came from a
German geologist, Hans Reck, shortly before the First World War.
Not only was Reck the first European to behold the Olduvai Gorge in
Africa's Great Rift Valley, his team was the first to recognise a hominin
fossil there.

As confirmation of Darwin's theory and for palaeontology more
generally, the Great Rift Valley is vital and famous in the study of
human evolution because of the fossil riches preserved by its unusual
geological properties. The term originally described a 3,700-mile
trench running from Lebanon to Mozambique, a fascinating geological formation that emerged from the splitting of the earth's crust.
However, most researchers today understand 'Great Rift Valley' to
refer to something smaller, to the part of East Africa where new tectonic plates are forming and literally beginning to tear the African
continent apart. To find such a place anywhere on earth is to find a
time-machine. Descending through the geological layers in the Rift
Valley is like travelling back through history to prehistory, a journey
of several million years. Even though the interpretations of finds in the
valley are often complicated by mixing and corruption of fossil sites

by tectonic upheaval, flooding, volcanic activity and so on, the Great Rift Valley has been and continues to be of inestimable importance for evolutionary theory.

When he was there in 1913 to study the earth's geological history and to excavate fossils, the twenty-seven-year-old Reck recognised this. And his work paid off. Near the end of three months of hard work, in the formidable East African equatorial heat, a crouching skeleton was discovered in one of the oldest layers of the gorge. Reck recognised that the remains he had discovered were of a Pleistocene *Homo sapiens* who probably had drowned there some 150,000 years ago.

The year was an ominous one, of course. Soon the 'War to End All Wars' began and palaeoanthropological research was suspended in order to carry out the sinister work of mass killing. For that reason, not much else was to come from Olduvai until the arrival of Louis Leakey more than twenty years later.

Leakey was a controversial researcher who energised palaeoanthropology in much the same way that Chomsky did linguistics and Einstein physics (though Leakey did not lead as a theoretician). He shook up his field by grandiose claims that attracted publicity both to the field and to Leakey himself. Along the way he and his family discovered some very important fossils in East Africa. Louis also fostered research on primates in their natural habitat, recruiting and encouraging researchers such as Jane Goodall (chimpanzees), Dian Fossey (gorillas) and Birutė Galdikas (orang-utans) to undertake their own field research.

After both advancing the earlier research of Hans Reck and eventually working alongside Reck himself, Leakey and his team discovered artefacts such as Olduwan and Acheulean tools, a skull of *Paranthropus boisei*, then called *Zinjanthropus*, and *Homo habilis*, among many others. Leakey and the headline-making publicity he received attracted many scientists to palaeoanthropology. Whatever his shortcomings, he earned his place as one of the innovators and founders of the field of palaeoanthropology.

More importantly, the findings of Leakey and other palaeoanthropologists have provided incredible insights into the evolution of our species. We now know that the human skeleton evolved over the last 7 million years or so, from the first likely hominins. Some of the features that distinguish us from other species include bipedalism, encephalisation, reduction of sexual dimorphism, hidden oestrus, greater vision and reduced sense of smell, smaller gut, loss of body hair, evolution of sweat glands, parabolic U-shaped dental arcade, development of a chin, styloid process (a slender piece of bone just behind the ear) and a descended larynx. These traits have become important to the classification and understanding of the place of different fossils in the hominin line.

One of the adaptations of human skeleta to the world around them came as evolution provided a novel form of locomotion. Humans are the only primates that walk upright. Other primates favour crawling or tree-swinging to get about. But to walk habitually (unlike a chimp, an orang-utan, or a bear that can walk upright only occasionally and for brief periods), our skeleta needed to change from the basic primate model to support this upright posture. One example of its many changes is found in the hole at the bottom of our skulls, called the foramen magnum. This is the aperture through which our spines connect with our brains. When this is found at the back of the underside of the cranium, we know that the creature did not walk upright regularly, because it would have been extremely uncomfortable. The spine would emerge nearly parallel to the ground for a creature on all fours, but awkwardly incline the head if the creature walked upright.

Another important milestone, the human head and brain, was achieved by a long process of encephalisation, the gradual process by which our brain cases got larger. Hominin brain case volume increased from about 450cm^3 for australopithecines to *sapiens'* 1,250cm^3. The heads of hominins show larger and larger brain cases until the appearance at *Homo sapiens* (*neanderthalensis* had even bigger brains than *sapiens*, averaging about 1,400cm^3 for males). Sapien skulls are large, rounded and delicate compared to the smaller brain cases and thicker skulls of our hominin ancestors. Gone are the special ridges at the top of hominin skulls to anchor muscles for chewing, along with the heavy brow ridges that perhaps shaded our eyes from the sun. In their place came a larger brain. And our heads developed accordingly, to give room and horsepower for thinking.[1]

Male and female bodies also grew more similar in size – that is, our sexual dimorphism was reduced. Although human males are roughly 15 per cent larger than human females on average, this size difference is smaller than that of any other primate species. The reduction of sexual dimorphism in the primate line has social implications. When males and females become more similar in size, this correlates, among primates, with pair-bonding, or monogamy. Male primates spend more time helping females feed and raise children. This is particularly important for human primates, since our children require a longer time to mature.

In some Western, industrialised cultures as much as one-third of a person's overall life expectancy is 'childhood' – the length of time required to reach autonomous adulthood. If males and females bond for life or simply in order to raise children, then the male will no longer need to battle other males for mating access. This reduces the pressure for males to have larger physical size, longer canines and other features for fighting. Battle is no longer necessary in order to pass our genes along to the next generation.

Along with bipedalism and reduced sexual dimorphism came a greater reliance on vision. Humans can see further than other primates, and most other creatures, which enables them to run faster towards a visible goal. Moreover, beginning with the arrival of *Homo erectus* humans acquired the capability of 'persistence hunting', running down game until it tires and the hunter kills it with a stone axe or club, or until it dies of exhaustion and overheating. Persistence hunting is seen even today in societies such as the Gê communities (Mebengokre, Kīsedje, Xerente, Xokleng and others) in the savannah regions of the Brazilian Xingu river basin.*

Evolution is also the ultimate economiser. With humans' greater dependence on vision came a loss of acuity and range in their sense of smell. If one portion of the brain gets bigger or better, another part very often grows smaller in the course of evolution. Here, the ability to smell degenerated as the vision region of human brains grew. Today the portion of the brain available to vision is roughly 20 per cent.

---

*Humans are able to run down game for several reasons. First, unlike any quadruped, humans are able to breathe hard while running. Second, humans' lack of fur, their perspiration, and their upright posture (with its greater surface area exposed to evaporation of perspiration) allow them to cool far more efficiently than quadrupeds. A human running after a horse, other things being equal, will eventually catch it.

(Fortunately, if someone is born blind, the vision region can be enlisted for other functions – evolution is often an efficient, no-waste process.)

Other changes to human physiology, not all with an immediately obvious intellectual benefit, might also have enhanced our species' intelligence. In the course of evolution, the length of the gut of hominins shrank. Intestines and digestive processes required fewer and fewer overall calories, enabling *Homo* bodies to shift more of their energy resources to their growing, ever-hungry brain with its expanding cranium. But natural selection does not receive all the credit for this change. Cultural innovation also played a role.[2] *Homo erectus* learned to control fire as long as one million years ago. As pre-*erectus* hominins learned to eat cooked food, the fats and proteins that they then ingested were much easier for their digestive system to break down. Whereas until this time, hominins, like other primates, needed larger guts in order to break down the large amounts of cellulose in their diet, as hominins learned to cook they were able to eat more meat and consequently able to consume more energy-rich food and to reduce significantly their dependence on uncooked plants that were much harder to digest. This fire-enabled dietary change facilitated natural selection's ability to produce larger brains in hominins, because their digestive organs required less energy and less space in the human body, while at the same making it possible to consume far more calories far more quickly (assuming the availability of meat). Cooking also altered our faces. It rendered the massive jaw muscles of pre-*Homo* hominins redundant and made our faces less prognathous. This in turn reduced the burden on the cranium to offer supporting structures, such as the mid-sagittal crest of the australopithecines, that arguably impeded the growth of the human brain case.

There are critics of this change-through-fire hypothesis. They hypothesise that *Homo erectus* was a scavenger and a hunter, finding rich sources of meat from carrion and fresh kills of its own long before controlled fire. Whatever the reasons, this reduction in gut size represents the movement to modern human anatomy. When encountered in the fossil record it is therefore a clue that the species represented by the particular fossil could be further along the evolutionary line to *Homo sapiens*.

Other important physiological changes needed for us to become modern humans included our upright posture and its by-products. As humans stood erect and walked habitually upright, their bodies became more efficient at thermal regulation. Moreover, since an upright body's surface areas are less exposed to direct sunlight than a quadruped's, hair became less necessary for humans. As a side benefit of shedding body hair, it became easier for humans to cool their bodies. They also evolved sweat glands in conjunction with the hair loss, making thermal regulation much more efficient. In hot, dry climates, the absence of hair and the production of perspiration allowed humans to cool off more quickly than many other animals. And sexual selection may have further sped up hair loss if people preferred less hirsute mates. This was all important to sustaining the human metabolic rate, so crucial for our intensely calorie-consuming brains.

Another characteristic of modern humans is their parabolic dental arcade. The evolution of human dentition has many causes and effects and dentition is important to fossil classification. *Homo* species' teeth shrank relative to their overall body size. Its canines in particular became smaller, which is important because this meant that *Homo* males no longer needed the larger teeth of other primates in order to fight for mating rights.

As the human dental arcade became more parabolic in shape, their faces came to possess more space for the articulation of different consonants and greater resonance for vowels, making a larger array of sounds available for human speech.

In order to summarise the output of human evolution in relation to other primates and to better understand the fossil record, a review of the primate phylogenetic tree on page 24 above is important.

Taking discoveries in order of the age of the fossils found, the first 'node' in the primate tree that links to humans (and chimps) is quite possibly *Sahelanthropus tchadensis*, literally Man of Chad of the Sahel area ('sahel' being cognate with Sahara, which gives name to what is today the largest desert on earth). *Sahelanthropus* is a potential direct link between humans and chimps but the more likely hypothesis is that it, like *Orrorin tugenensis* and *Ardipithecus*, was one of the first hominins as in the lower right portion of Figure 4. The repetition of the names above and below the split indicates the two major hypotheses regarding these hominin fossils. It lived some 7 million years ago. Though we have

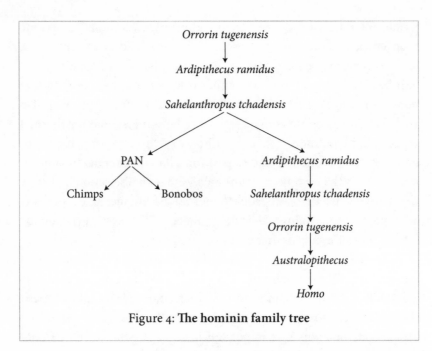

Figure 4: **The hominin family tree**

only parts of the cranium, mandibles and teeth of *Sahelanthropus*, it is important to the fossil record of the evolution of *Homo sapiens* because it represents several distinct but equally important possible clues to human evolution. It could even be the 'node' in the phylogenetic tree from which chimpanzees and humans split (Figure 4).

Before concluding the discussion of the rise of hominins, it would be useful to compare what is emerging about the cognitive and communicational abilities of other great apes in order to more effectively reflect on the nature of the different evolutionary paths – if any – to language.

Mammals are the most intelligent creatures of the animal kingdom. Primates are the most intelligent mammals and humans are the most intelligent primates. Therefore, humans are the most intelligent animals. This may not be saying much. After all, our intelligence is the reason we murder one another and fight wars. Our brains are a mixed blessing. Jellyfish get along quite nicely without brains.

Nevertheless, human language is possible only because of this greater intelligence. Humans are the only creatures known that use symbols and cooperate to communicate more effectively. And unlike other animals, when humans communicate they almost never say all that they think, leaving their hearer to infer meaning.

There are some prerequisites, or what are often called 'platforms', needed for language.[3] Two of those are culture and 'theory of mind' – an awareness that all people share cognitive abilities. Culture is an important topic, but right now it is worth discussing the theory of mind, because this also helps gauge an area of cognition where humans are regularly claimed to have something that no other animal has – the ability to 'read' the minds of others. Although actually being able to see or hear the thoughts of others is science fiction, there is some truth to the idea that humans can guess what others are thinking and then use this knowledge as a key to communication.

To cognitive scientists such as Robert Lurz, mind-reading is 'the ability to attribute mental states, such as beliefs, intentions and perceptual experiences, to others by the decidedly mundane and indirect means of observing their behaviours within environmental contexts'.[4] An example of this might be to see a man with two bags of groceries standing outside the entrance to a house feeling around with his free hand in his pockets. A person with a knowledge of locks and keys and the custom of locking one's house should be able to guess that the person is searching for his keys and that he plans to unlock and then enter the house. Even for this seemingly simple scenario, there is a huge amount of cultural knowledge being drawn from. Amazonians who lack locks and keys might not have a clue as to what the man has his hands in his pockets for. And yet all humans will most likely recognise that the man has an intent, a purpose or a goal; that his actions are not random. That's because all humans have very similar brains, which is in essence what the theory of mind is. Folks with autistic spectrum disorder might not understand this, because there is reason to believe that some forms of this set of ailments are caused by a lack of this kind of awareness.

Language works only because people believe other people think enough like they do to understand what they want to tell them. When one says what one is thinking, they do so believing that their hearer will be able to understand, infer conclusions about and match our words to their own experiences. Therefore, the question that arises is whether humans alone in the animal kingdom have this ability. If other creatures possess it, what does that mean for their systems of communication, their cognition and the evolution of human language?

Studying animal behaviour (just like studying the cognitive abilities of human infants before they can speak) is extremely hard because of the danger of overinterpretation. To take an example from my Amazonian field research, consider the Amazonian horsefly. The bite of this nasty little creature hurts more than most because they (the females only) suck blood by lacerating the skin. What's worse is that the locations of the bites itch for a good long time afterwards. Hiking through the jungle is almost always rewarded with multiple bites from these pests, along with their partners in crime, mosquitoes, wasps and smaller species of blood-sucking flies. One thing about horseflies, though, is that they seem to know where you are not looking!

On a certain level, it often seems as though Amazonian horseflies must have minds that can figure out human behaviour. While it is true that they seem to use a strategy for choosing a location on the body (clothes are no impediment as they can easily bite through denim jeans and cotton T-shirts) based on an interpretation of other animals' behaviour, should one then say that horseflies have a plan for sucking blood that is based on interpretation of their victim's perceptions? Doubtful.

An alternative explanation could simply be that the flies are genetically programmed to bite the relatively darker areas of a victim – the shaded part of their appendages. A shaded part of your body will also be one in which the visibility will be much reduced. People often anthropomorphise and interpret as cognitively designed actions what are in all likelihood physically determined.

Getting back to primates and animals more generally; there are many rigorous studies that avoid overinterpreting animals' behaviours.[5] One of the most problematic issues in the lengthy conversation in science about whether animals have cognitive abilities in any way similar to those of humans is the profoundly circular assumption that cognition requires language, human language at that, and that therefore animals cannot have cognition because they lack language. This is simply declaring by fiat that humans alone have cognition, before research has been conducted. Such ideas are misguided by their anthropocentric framing of the questions.

These views derive from the work of René Descartes in the

seventeenth century, who believed that only humans 'think therefore they are'. Descartes's views arguably set back studies of human cognition because they discouraged comparative evolutionary studies of mentality. They also affected non-human studies by simply declaring that animals lacked mental lives.

In Descartes's view, non-humans possess no consciousness, no thought and no feelings. Additionally, his view that human minds are disconnected from bodily experience led instinctively to his linguistic-based theory of cognition, namely that only language users think.

But as philosopher Paul Churchland aptly puts it: 'Among many other defects, it [the account that only humans think because thinking requires language] denies any theoretical understanding whatever to nonhuman animals, since they do not traffic in sentential or propositional attitudes.'[6]

Any view of cognition that ignores non-human animals ignores evolution. Whether we are talking about the nature of ineffable knowledge or any other kind of cognitive or physical capacity, our account must be informed by and be applicable to comparative biology if it is to have any explanatory adequacy. Animal cognition helps understand the importance of evolutionary theory and comparative biology in the understanding of our own cognition. It also allows for tremendous insight into how the bodies of both humans and other animals are causally implicated in their cognition.

The main problem with disregarding animal cognition is that, in doing so, we are essentially disregarding what cognition might have been like among our ancestors *before* they got language. Their prelinguistic state was the cognitive foundation that language emerged from. If there is no cognition before language, à la Descartes and many others, the problem of understanding how language evolved becomes intractable.

Of course, there are those who claim that language did not evolve gradually, so we wouldn't expect to find its roots in any other species. According to such researchers, the grammatical core of language 'popped' into being via a mutation, bringing forth a linguistic Prometheus whose X-Men genes spread quickly throughout the entire species.

On the other hand, there are those who work experimentally to address the question of whether primates have beliefs and desires and whether other primates are capable of 'mind-reading'. For both questions the evidence so far answers tentatively 'yes' – there does seem to be some form of these abilities in other primates.

So humans may not be alone in the world of thinking and interpreting others. But if other primates, such as chimpanzees with their 275–450cm^3 brains, are capable of reading the intentions of other creatures, as well as holding beliefs and desires, then surely the 500cm^3-brained primates of the genus *Australopithecus* or the 950–1,400cm^3-brain-sized species of *Homo* had even more well-developed powers of cognition and social understanding.

Animals and fossils strongly support the idea that humans got their unique abilities by baby steps. And our debt for this knowledge goes back to the fossil hunters. The painstaking work of collecting fossils and attempting to piece together cultural and anatomical evidence for the origins of our species takes physical fortitude – to withstand the heat, sweat, remoteness and occasional danger of palaeontological field research. It is a cut-throat, competitive enterprise at times, with mudslinging from all sides.

But in spite of the hard, painstaking field research of palaeontologists, on 1 January 1987 an article appeared in the journal *Nature* which threatened to wrest all the glory, power and science from the palaeoanthropologists and transfer it to lab-coat-wearing geneticists. The paper, 'Mitochondrial DNA and Human Evolution', co-authored by Rebecca L. Cann, Mark Stoneking and Allan Wilson, argued that genetic evidence clearly established that the DNA of all current *Homo sapiens* traces back to a single female's mitochondrial DNA about 200,000 years ago in Africa.

This was a bombshell. Could it really be that three people in a comfortable laboratory put an end to the controversy surrounding the 'recent out of Africa' vs 'multiregional' hypotheses? To review, the former claimed that *Homo sapiens* originated in Africa and migrated out, replacing other *Homo* species across the globe. The latter suggested that all modern humans evolved in separate lineages from the various sites of *Homo erectus* around the world.

It turned out that the multiregional hypothesis was shown to be largely incorrect. When it first became public, the 'Mitochondrial Eve' theory was met by criticism from the proponents of the multiregional hypothesis, among others. But it has held up well to scrutiny and is now accepted widely by palaeoanthropologists, biologists and geneticists. The lab workers beat the field workers on this one.

Before one can grasp the significance of Mitochondrial Eve for language origins, however, it is necessary to review the science behind the conclusions. This is the theory that underlies the notion of a molecular clock on which the Mitochondrial Eve story is based. Originating sometime in the early 1960s and first published in a paper by Linus Pauling and Emile Zuckerkandl, the molecular clock idea came about after noticing that changes in amino acids across species are temporally constant. Thus, knowing the differences in amino acids between two species can tell when these species split from a constant ancestor.

As with most scientific discoveries, several people soon added to these ideas. Then in 1968 Motoo Kimura published a now famous article, 'Evolutionary Rate at the Molecular Level', in *Nature*. Kimura's paper laid out the basic ideas of a 'neutral theory of molecular evolution'. The neutral theory here is non-Darwinian, meaning that, rather than natural selection, Kimura placed the responsibility for most evolutionary change on genetic drift produced by random, neutral variations in organisms. Since these changes do not affect the survivability of an organism, it is able to pass on its genes normally to viable and fertile offspring.

Applying the molecular clock to mitochondrial DNA collected from humans around the world led to the proposal that all living *Homo sapiens* come from a single woman (called 'Lucky Woman' or 'Mitochondrial Eve') in Africa, about 200 millennia ago. In other words, only one woman from the past produced an unbroken line of daughters up until the present, thus transmitting her mitochondrial DNA to all living humans.

The genus *Homo* thus arose in Mother Africa. But if life was so good in Africa, why, when and how did our *Homo* ancestors leave there?

3

The Hominins Depart

We travel, some of us forever, to seek other states, other lives, other souls.

Anaïs Nin

THE GREATEST HUNTER. The greatest communicator. The most intrepid traveller. Perhaps the greatest distance runner on earth, *Homo erectus* was the unsurpassed marvel of its time. No other creature has ever contrasted more starkly with all the animals that had ever lived. *Neanderthalensis* and *sapiens* were born from and first lived in the shadow of *erectus*. We were not new. They were. *Sapiens* are just the improved model of *Homo*. *Erectus* was the first to journey. They were the original imagination-motivated travellers.

Of course, travel itself did not begin with *Homo*. Many species move from one environment to another. Migration sets up competition with the local species. The genus *Homo* is no different. Yet so early did *Homo* begin to travel that, although they originated in Africa, their first fossils were found not there but in Asia – in Indonesia and China. Later, the fossils of other *Homos* were discovered in Europe – in Spain, France and Germany. How did these fossils come to be in these places? It seems like a nearly impossible task for humans to actually walk around the world. But they did. And for *Homo*, with its nearly unprecedented endurance, the trip wasn't as hard as it sounds.

Initially *erectus* and other *Homo* species were hunters and gatherers. As such, they needed to move frequently as they exhausted the edible flora and fauna of a given region in a relatively short period of time. Hunter-gatherers usually move a bit further each day from their original village. They may sometimes return to an established camp, but as food becomes ever scarcer in the area surrounding the original village, hunter-gatherers move to establish new settlements closer to their fresher sources of proteins and plant foods.

The average forager travels just over nine miles (fifteen kilometres) per day. Assume that they move communities around four times per year and that each new village is a day's foraging from the last village. That is thirty-seven miles (sixty kilometres) per year. How long, at that rate, would it take an *erectus* community to travel from Africa to Beijing or Indonesia, both locations where *erectus* fossils were found? Well, if one divides 10,000 kilometres (roughly the distance from East Africa to where *erectus* fossils have been found in China) by sixty (the number of kilometres *erectus* would, under my extremely conservative calculation, move in a year), then it would take only 167 years for *erectus* to traverse Eurasia, moving at a normal pace. But if *erectus* populations had other reasons to travel, such as to escape hostile neighbours or climatic events such as flood or drought, they might have moved even more quickly, potentially reducing the time needed in the extreme case to as little as a year. Likewise, if they moved more slowly due to, say, the discovery of rich food supplies in a place along their route, then a larger period of time would elapse. It was, in any case, easily within the grasp of *erectus* to settle large regions of the world within only a thousand years, a trivial amount of time from an evolutionary perspective.

In the course of their earliest journeys, *erectus* populations would never have encountered any other humans. They were the first to arrive at every destination. They had all the natural resources of the world before them, with all the land they saw at their disposal. *Homo erectus* men and women were the greatest and most fearless pioneers of our species.

Throughout modern history there have always been refugees and migrants, people fleeing wars or famine, looking for a better life, or just satisfying their own wanderlust. The genus *Homo* – both *sapiens* and their ancestors – has always been the wandering kind. But unlike any other species, *Homo* species probably all talked about their migrations. And their conversations about travel made the trips more enjoyable. Humans don't migrate like other mammals. We *sapiens* plan our trips, review them, celebrate them and lament them. And *erectus* seems no different, from what we can tell from the fossil record.

This conscious movement to the unknown was but one of the cognitive capabilities that emerged in humans on their long path to becoming sapient. Their new-found consciousness was a state of mind that exceeded mere animal awareness. Gradually our ancestors' consciousness came to include self-referential reflection: not only did their thinking include 'I am aware of x' but also 'I am aware that I am aware of x'. This is 'conscious consciousness' and it would have facilitated their travel as well as their thinking about the symbols that were already emerging from the growing complexity of their cultures. *Homo* species in all probability began their perambulation with self-conscious purpose. *Erectus* would have been the first creature in history to be self-conscious. And the first to imagine. (Imagination is the knowing consideration of 'what is not but could be'.)

Homo erectus would have initiated the sharing of values that is unique to human societies. Social roles began to emerge as communities discovered that different community members were better at some things than others. These ancestors began to remember and to organise the knowledge they were gleaning from the world around them and from each other. And they taught their children these things. This is inferable from the ever-improving tools, homes, villages and societal organisation that have been found in the fossil record. Humans were getting smarter. They were becoming cultured. They had crossed the communication-language threshold.

There were changes in *Homo erectus* that no other species in the history of the planet had undergone. *Erectus*'s achievement of self-conscious cognition quite possibly enabled them to (eventually) talk about, characterise, contextualise and classify their emotions – love, hate, fear, lust, loneliness and happiness. Our ancestors in all likelihood also began *to keep track of* their kith and kin on their travels. And this growing knowledge, as it emerged from their evolving culture and travels, would have eventually *required* them to invent language of some sort (with their relatively enormous brains). And increasing culture placed evolutionary pressure on *erectus* to evolve ever more effective and efficient linguistic abilities, accompanied gradually by the brains, bodies and vocal apparatuses necessary to exploit those abilities fully. At the same time, at the interstices of culture and language, *erectus* would have been developing what can be referred to as 'dark matter of the mind' – tacit, structured knowledge, prioritised values and social roles. Dark matter is crucial to the interpretation and arrangement of human apperceptions (experiences that affect our development, stored in our unconscious that create individual psychologies).

The emerging psychologies of *Homo erectus* would have interacted with their community to produce culture. *Erectus* followed the more hospitable and passable swathes of East Africa until they emerged from the continent, initially in the Levant and then on across Eurasia and to islands across the sea. They were the Argonauts of the Pleistocene. And as they arrived they were more sophisticated than when they left. They were also better fed.

Hunting prowess and the advantages of a meat diet fuelled *erectus* travel. The hunt provides much more than fat and flesh. Hunter-gatherers eat the skin, bones and offal. They consume nearly the entire animal head to toe. They eat the bones by splitting them and scraping out the marrow. Then they shave bone fragments off that are so thin that they can be eaten without difficulty or they are boiled and consumed. And then, after eating this large quantity of animal protein and calcium, they can rest for a day or two before needing to hunt again, depending on the size of the animal killed. Like modern hunter-gatherers, *erectus* also controlled fire. It not only killed better than other creatures, it ate better and healthier. And it transformed its body and its brain. Fire would have been tremendously helpful during the trip, allowing them to travel further in a day, chat around the campfire at night and build strong ties of community.

Anyone who has accompanied hunter-gatherers as they pursue game for several miles without rest knows the joy of talking about the hunt at night around the fire. Sometimes they sleep near their kill because they are too tired to make their way back to their village. Then, the next day, after again eating very well, *erectus* bands might have bundled up their leftovers with vines or have simply tossed large portions of the animal, such as a leg or hindquarter, across their shoulders as they headed back to their families. If their families were already with them, perhaps they remained a bit longer near the latest kill site. They may have tarried a day or so, then explored further around their new campsite, after consuming all of the meat from the previous day. Perhaps they relocated their village to the site where they successfully killed the game, especially if it had more abundant edible plants or more game.

How easy was it to travel around in Africa at the time of *Homo erectus*? During this period, roughly 2 million years ago, Africa was climatologically very different from today. The so-called 'Sahara pump' was active then. The current Sahara desert was then non-existent. Instead, all of North Africa was covered in lush forests that stretched across the Middle East and on through Asia. Flora and fauna were rich throughout large swathes of the world that are today barren deserts. This ecological-climatological fecundity dramatically contrasts with today's North African climate and it clearly supported the exploration and nomadism of *Homo erectus*. Major changes in human genes were also happening at this time, changes that I suspect would have facilitated the expansion of *Homo erectus*'s geographical range, even without the Saharan pump.

Erectus was truly marvellous. But in spite of the admiration they richly deserve, these people were not the equals of *Homo sapiens*, not even of *neanderthalensis*. They were simply the first habitually upright hominins and the first humans. They were the first interpreters of their own visions, as they were the first bearers of culture and the earliest storytellers of our planet. They were the progenitors of both *neanderthalensis* and *sapiens*. Their skulls and bodies were becoming more modern, though they still had prognathous jaws, making them look somewhat ape-like (Figure 5).

What is really known about *Homo erectus*? Did they really have language or were they just grunting cavemen? Like most areas of human endeavour, ignorance outstrips knowledge. There is much to learn about these ancestors before they and their role in the evolution of the genus *Homo* is fully understood.

On the other hand, less is known than one would like about a lot of things that researchers hazard informed hypotheses about. So this should not stay anyone from considering ideas that are supported, however shakily at present, by the facts. For whatever reason, a subset of the *erectus* people decided to leave Africa about 1.8 million years ago. Their travels began only a couple of hundred thousand years after they first appeared. Not long after that (in geological time, very quickly: only about 200,000 years), confirmed evidence shows them in South Africa, the Middle East, modern-day Georgia, Europe, China and Java.

Homo erectus evolved in the Pleistocene from australopithecines.

Figure 5: *Homo erectus* (artist's impression)

Their bodies got bigger. Their brains got bigger. Their societies grew more complex. Their technology developed quickly. Why did this transformation emerge during the Pleistocene? Why not later or earlier? Is this a mere coincidence? Most think not. The Pleistocene posed the problem of survival as it had never been posed before for hominins. Its rapid climate variations, advancing and receding glaciers, changes in flora and fauna were among the challenges it forced hominins to adapt to.

According to some classifications, there were, soon after and before *erectus*, other species of *Homo* co-existing or existing in close succession – *Homo habilis, Homo ergaster, Homo heidelbergensis, Homo rudolfensis* among others. But, again, most of these various species of *Homo* are ignored here, with the focus kept on *Homo erectus, Homo neanderthalensis* and *Homo sapiens*. Most other *Homo* species are murky, maybe nothing more than variants of *Homo erectus*. However, the story of human language evolution changes in no significant way, whether *erectus* and *ergaster* were the same or different species.

Remaining with this simplified inventory of species of *Homo*, *Homo erectus* was in all probability well on the way to inventing language by roughly 1.9 million years ago. They used tools. This brain size resulted from many pressures – the advantages of improving tools, the need for better communication to keep track of social relationships, travel and the need to cope with a rapidly changing environment. As the climate became more arid and colder in East Africa, *erectus* trekked to the south of the continent.

It is no coincidence that the greatest changes and innovations in human physiology, cognition, sociality, communication, technology and culture (dwarfing any of today's inventions and developments) occurred during the Pleistocene. Glacial sheets covered the northern hemisphere many times during this period. Some pre-*Homo* hominids adapted physiologically to the greater aridity of the environment. *Paranthropus*, a genus of 'robust' australopithecines contemporary with *erectus*, grew bigger teeth, with thicker enamel in order to eat seeds that became larger and harder to crack during this time.

But *erectus* relied on culture to solve problems posed by their volatile environment. Instead of their teeth, *erectus* used rocks for cracking seeds, thereby adding cultural pressure to evolve more and more intelligence to make better tools. This was what I call the 'first cultural revolution', where our ancestors changed culturally in order to answer the ever new challenges of their environment.

It was during this time, over 2 million years ago, that a common set of stone tools, the Olduwan tool kit (flaked rock tools named for the site of their earliest discovery, by the Leakeys in the Olduvai Gorge), first appeared in the archaeological record. This tool assemblage may (or may not) have been used by australopithecines initially. Similar but non-identical tools can even be used by chimpanzees and other non-human apes (in ways quite distinct from human usage), though a great deal of practice is required. Regardless of its first users, however, this tool kit was widely employed by *erectus* and other *Homo* species. The arrival of tools signifies that culture is beginning. And the birth of culture has implications for language evolution and physiological adaptations during this time.

Figure 6: **Olduwan tool kit**

The Olduwan tool kit shown in Figure 6 was made by a process of flaking, which began rather simply but eventually led to quite complex skills.

Though occasionally it is said that animals such as otters, chimps and orang-utans have 'culture' based on their use of tools, real culture is far more than merely this. Likewise, culture is more than the *transmission* of tool technology or other knowledge from one generation to another by imitation or overt teaching. Culture attaches values, knowledge structures and social roles to humans and their creations. This means that even tools have meaning. Because of this they bring to mind for the member of a culture the tasks they perform, even when those tasks are not currently under way. A stone axe on the ground can elicit memories of the times they have been carried on trips. It can also bring to mind a previous user.

Cultural implications thus surpass mere tool use in cognitive complexity. When an orang-utan uses a stick as a spear to catch fish in Borneo, or a chimp uses a chair to climb up over a fence, or an otter uses a stone to open a shellfish – even if their offspring learn to use these from them – this does not mean that they possess culture. They are using (perhaps even transmitting) tools in the absence of culture. Impressive as tool use is, culture goes beyond this by *contextualising* artefacts. This is what enables tools originating in a particular culture to evoke meanings even when they are not being used. A member of a culture that uses shovels or scissors knows what shovels or scissors are for even in the absence of their associated activities. The tools alone will bring those activities to mind. Outside of culture, tools evoke no abstract connections to values, social roles, or knowledge structures. One can tell the difference only by examining the evidence that tools emerge from a system rather than from a one-off or idiosyncratic invention, as perhaps a chance usage by a single family or individual. We might question whether the tool plays a part in distinguishing social roles or relative to other tools, or attempt to determine its value relative to other tools of the culture. Is it used only by some people or by everyone? Does it have a specialised purpose?

Other evidence of incipient culture among *Homo erectus* populations is one previously mentioned, namely that *erectus* adapted physiologically to a relatively rare way of life among animals – pair-bonding – a social structure in which males and females mate long term wherein the male feeds and protects the female and their offspring in exchange for near-exclusive sexual access. Pair-bonding is inferred not only from the archaeological record of *erectus* villages but also from smaller *erectus* canine teeth and reduced sexual dimorphism between males and females. Pair-bonding plus tools is evidence for family units and cooperation.

This view of human cooperation in *erectus* is strongly supported by the archaeological record. As *erectus* wandered through the Levant, near the Jordan between the Dead Sea to the south and the Hula Valley to the north, they came to stop at the site known today as Gesher Benot Ya'aqov. At this site, going back at least 790,000 years, there is evidence for Acheulean tools, Levallois tools, evidence of controlled fire, organised village life, huts that housed socially specialised tasks of different kinds and other evidence of culture among *Homo erectus*. *Erectus* may have stopped here on the way out of Africa.

Erectus technology was impressive. They built villages that manifested what almost appears to be central planning, or at least gradual construction under social guidance, as in Gesher Benot Ya'aqov. This is clear evidence of cultural values, organised knowledge and social roles. But such villages are just one example of *erectus*'s technological and organisational innovation.

Another may be seen in the routes they followed. As specialists have mapped out the travels of *Homo erectus* around the world an interesting observation comes to light – *erectus* seems to have deliberately travelled to geologically unstable areas. *Erectus* followed a route known as the Plio-Pleistocene Tethys (the former coasts of an even more ancient ocean), which provided a natural geographic path, along with geological instability.*

Whether it ultimately turns out to be correct or not, the idea that geology played a major role in the routes of the migration of *Homo erectus*, rather than simply random wanderings about the earth, offers clues to the species' thought processes. All humans make decisions and they marshal evidence for those decisions. It would be extremely surprising if *Homo erectus* did not have reasons for going left or going right as they travelled. Though culture also played a role, the Plio-Pleistocene Tethys offers a simple possibility – namely that *erectus* followed the 'lie

---

*In a web-based discussion (http://www.athenapub.com/13sunda.htm) authors Roy Larick of the Shore Cultural Centre, Euclid, Ohio, Russell L. Ciochon of the Department of Anthropology, University of Iowa, and Yahdi Zaim of the Department of Geology, Institute of Technology, Bandung, Indonesia, claim that:

Fossils representing very early *Homo erectus* populations are now known from the highland Rift Valley of East Africa, the Caucasus Mountains that mediate southeast Europe and southwest Asia, and from the intensely volcanic slopes of the Sunda subduction zone. Circum-Mediterranean archaeological sites representing these groups may be present in northern Algeria (Ain Hanech), Andalusian Spain (Orce), and the Negev (Erq el Amar). Late Olduvai subchron archaeological sites are also known on the Himalayan fore slope (Riwat, Pakistan), and in southern China (Longgupo). The Plio-Pleistocene carnivores associated with humans are also known from Greece (Mygdonia Basin).

The commonalties among these sites call for a new interpretation of early *Homo erectus*. All these sites fall into the transcontinental Tethys geotectonic corridor, the grand suture at the southern margin of the Eurasian continental plate with southward extensions into the East African Rift and the Sunda subduction zone. A global time marker immediately precedes and overlaps with all sites, the Olduvai subchron (1.96 to 1.79 mya [million years ago]). With the corridor and the subchron, we can begin to talk about *Homo erectus* biogeography as neither African nor East Asian, but as Plio-Pleistocene Tethys.

of the land'. There were geological conditions favourable to the route that *erectus* chose. If this is correct it is an interesting finding. However, before we can definitively interpret the routes of *erectus* as based on culture and cognition vs simple hunting like any other animal, we would need to compare their routes of migration to those of other animals that left Africa. And then we'd have to determine whether *erectus* was simply following other animals or whether they were being guided by hunger rather than cultural values or knowledge structures.

However, the possibility that *erectus* was travelling based at least partly on culturally guided or otherwise intelligent decision-making is supported by other finds in the record. One of the greatest surprises in archaeological history – and there have been many – was the discovery of Acheulean tools on the Indonesian island of Flores in 2004. This find was preceded somewhat by a discovery in 1957 by Theodor Verhoeven, a Dutch archaeologist and missionary, of bones of Stegodontidae, an extinct family of Proboscidea (relatives of mastodons, mammoths and elephants), on the same island. The stegodonts, like modern elephants, were very good swimmers. Elephants have been observed to swim for as long as forty-eight hours, in a herd, across African lakes. They are known to have swum as far as thirty miles (forty-eight kilometres) at sea (which is further than the distance to Flores would have been 750,000 years ago).

Flores sits among the lesser Sunda Islands of eastern Indonesia. The fifteen-mile (twenty-four-kilometre) strait separating Flores from the closest land, the source of the Stegodontidae, would not have presented a great swimming challenge to the large mammals, who pursued float-ing plants across the strait. But the tools later discovered near charred bones of these creatures do present an enigma. How did they get there? These tools are nearly 800,000 years old. And there is no period during which the island was connected to any other land. It has always been isolated by deep water. *Erectus* somehow got to Flores. How?

Unlike the Stegodontidae, they could not have swum there. Even had they spotted the island on the horizon and decided to visit, the currents would have made swimming there impossible. The greatest waterflow in the world is known as the 'Pacific Throughflow', and it flows around the islands of Indonesia, including Flores. These currents would defeat all but the most elite athletes. Yet there is evidence of a relatively large *erectus* population on the island. A founding population would need to include a minimum of fifty individuals. And it is unlikely that they all set out paddling logs or attempting to swim across treacherous currents, even though they may have witnessed stegodonts doing such a thing. They must have had a motive to go, certain that there would be plenty of food there.

The idea that a founding population crossed the straits piecemeal, without planning, is implausible – fifty or more 'shipwrecks' as it were, within a short time, where everyone survived. They would have had to arrive during a short period to guarantee survival and this would have required an unfeasible amount of coincidence. It is, of course, possible that a flotilla of logs was launched, of which fifty or more made it to the island. But, while that would not lessen the intent and adventure of *erectus* in crossing to Flores, it would provide a poor explanation for the settlements on Socotra and other islands described below, an island out of sight, requiring a sense of imagination and exploration for a large *erectus* population to arrive within a time frame short enough to guarantee their survival. Moreover, archaeologist Robert Bednarik and others have provided extensive and convincing evidence that *Homo erectus* built watercraft and crossed the sea at various times in the lower Palaeolithic era, around 800,000 years ago (and three-quarters of a million years before *Homo sapiens* made sea crossings). Bednarik has even built and sailed replicas of the kinds of bamboo rafts that he believes *erectus* would have constructed, fabricating water containers from bamboo and using techniques that would have been within the reach of *Homo erectus*.

Many archaeologists have provided evidence of *erectus* technology that, while not surprising for *sapiens*, force the reconsideration of the common view that *Homo erectus* could do little more than grunt to communicate and had no actual words. Further examples of *erectus* technology and art include decorations, bone tools, stone tools, evidence of adding colour to art, wooden artefacts, backed knives, burins (stone chisels) and protoiconic palaeoart.*

---

*Backed knives 'were made by steeply trimming one edge of a blade by pressure flaking. This design allowed the user to apply pressure against the blunted edge with an index finger for cutting with the opposite sharp edge. Experiments have shown

Given all of this evidence, it is nearly certain that *erectus* had developed culture. But 'culture', once again, means more than that they built tools or that they passed down the knowledge of how to build and use these tools to subsequent *erectus* generations. Culture entails symbolic reasoning and projecting meaning on to the world, meaning that is not about things as they are, but as they are interpreted, used and perceived by members of the community that uses them. Culture transforms 'things' into symbols and meaning. And if *erectus* had symbols, it had language.

The case for *erectus* culture is further strengthened when one learns that Flores is not the only island to which *erectus* voyaged. And although there are no remains of actual million-year-old wooden or bamboo boats that they might have employed, evidence that they inhabited isolated islands neither accessible by swimming nor visible from shore suggests very strongly that they intentionally journeyed miles across the open sea. This conclusion seems warranted, in spite of the fact that the oldest boats we have physical evidence of are dugout canoes from the Upper Palaeolithic, only a few thousand years old.

As recently as 2008, Russian researchers found very primitive stone tools on the isolated island of Socotra, more than 150 miles (240 kilometres) off the Horn of Africa and 240 miles (400 kilometres) off the coast of Yemen. And the timeline is roughly the same as it was for Flores – these discoveries are estimated to be from 500,000 to 1 million years old.

One can imagine the inspiration for the voyage to Flores – witnessing a herd of Stegodontidae swimming there. But this cannot explain the innovation, confrontation of the unknown and abstract thinking that were manifested by the *Homo erectus* population that sailed to Socotra, Crete, Flores and other islands. Indeed, on that voyage they seem to have been *exploring*, which requires a form of abstract thinking that goes beyond the here and now, the observable, to the *imagined*. And evidence of imagination is evidence for abstract thinking. Taken together, the currents *erectus* had to overcome to reach Flores and the challenge of the unknown on the voyage to Socotra establish clearly

---

that a backed knife made of stone can skin an animal about as fast as a steel knife.'
(www.lithiccastinglab.com/gallery-pages/aurignacbackedknifeag7large.htm)

that *erectus* cooperated for a common goal, utilising innovative technology. Such accomplishments imply the ability to communicate at a level more advanced than any creature until that time.

Of course, it is possible that *erectus* never intentionally sailed, but that they built rafts for fishing close to shore and were blown off-course to islands (or death) in the open sea. This likely happened at times. Modern sailors suffer the same fate occasionally. Yet even this possibility would be evidence that *Homo erectus* had enough language to build rafts. But this 'blown off course' suggestion fails to explain the settlements we see on various islands, from the Sea of Flores and the Gulf of Aden. For each viable settlement and subsequent cultural development, at least 40–50 *Homo erectus* men, women and children would have had to have arrived almost simultaneously.

But what kind of language did *erectus* speak? What minimal form of communication would it have needed? The answer seems to be something like what I refer to as a 'G_1 language'. This is a language in which symbols (words or gestures) are ordered in a conventional way when spoken (such as subject-verb-object, as in 'John saw Mary'), although, somewhat contradictorily, the interpretation of the symbols in this agreed-upon order can be very loose. Thus, 'Mary hit John' might mean in the first instance that Mary hit John, but might have other meanings available according to the context, such as 'Mary was hit by John' or 'Mary bumped into John' and so on. The context in which the words are uttered as well as what the speaker and hearer know about Mary and John, in conjunction with general cultural knowledge and the agreed-upon word order, will determine the interpretation intended. A G_1 language is nevertheless a real language. It is not some 'protolanguage' (qualitatively different from a 'real' language). Such a language can actually express everything needed by a particular culture and is 'expandable' to fit additional needs if the culture becomes more complicated. Think again of examples like 'No shirt. No shoes. No service.' This can mean quite a few things, but members of American culture at least interpret the phrase as an admonition from a business establishment, even though there is little in the words themselves that indicates such a thing. Culture serves as a filter on what the meaning is. Grammar is another partial filter. So in this case, 'Mary hit John' might mean that Mary hit or bumped into John, but it would be harder for it to mean that John bumped into Mary because of the word order imposed by the grammar, which acts as a (weak) filter on the possible meanings of the sentence. Whenever the grammatical filter is less fine-meshed, as in a G_1 language, the role of culture in aiding the meaning becomes even greater, though it is always present in all languages.

The archaeological evidence leads to the conclusion that *Homo erectus* possessed creative thought and culture. In other words, in spite of scepticism from some researchers, *erectus* spoke, was creative, and organised its communities by principles of culture. The cultural evidence is otherwise inexplicable. *Erectus* were seafarers and manufacturers not only of technologically interesting hand tools, from Lower Palaeolithic Olduwan tools to Upper Palaeolithic Mousterian tools, but also vessels able to cross large bodies of water. *Erectus* communities, such as the one Gesher Benot Ya'aqov, developed cultural specialisation of tasks. And *erectus* controlled fire, as evidence from several *erectus* sites suggests.

Once again, though, *erectus*'s speech and language may have differed significantly from those of modern humans, yet *erectus* languages nevertheless would have been full languages. So long as they possessed symbols, ordering of the symbols and meanings partially determined by those components in conjunction with context they had language. And it seems clear for various reasons that they would not only have spoken their language but also have used gestures as aids to communication. Neither gestural languages nor music nor controlled use of pitch (as in singing) would have come first (see chapter 10 below). Simple (G_1) languages emerged with grammar, which, accompanied by pitch modulation and gestures, produced the most effective communication system the world had ever seen. This is the minimum form of language possible.

Erectus speech, however it sounded, is an important but secondary question. *Homo sapiens*' still bigger brains, longer experience with language, a more developed vocal apparatus and so on, give us huge advantages. They mean that *sapiens*' languages are more advanced, in the sense of having larger vocabularies and probably more complex (hierarchical and or recursive) syntax. Nevertheless, the upshot is that there is no need to suppose that *erectus* spoke a subhuman 'protolanguage'.

A protolanguage by definition is not a fully developed human language, but rather merely a 'good enough' system for very rudimentary communication. But the kind of language that *erectus* would have used would have been good enough not only for *erectus* but also for modern *sapiens*, depending on the needs of individual cultures, because a G_1 language can communicate almost as well as a G_3 language.

Erectus travelled almost the entire world, though based on current evidence never made it to America, Australia or New Zealand. But they made it to many other places. Here is a brief summary of *erectus* sites and time ranges:

Middle East
Gesher Benot Ya'aqov (790 thousand years ago)
Erq al-Ahmar (1.95 million years ago)
Ubeidya (1.4 million years ago)
Bizat Ruhama (1.96 million years ago)

Italy
Pirro Nord (1.6 million years ago)

Turkey
Dursunlu (before 1 million years ago)

Iran
Kashafrud (before 1 million years ago)

Pakistan
Riwat (before 1 million years ago)
Pabbi Hills (before 1 million years ago)

Georgia (before 1 million years ago)

Spain (before 1 million years ago)

Indonesia (around 1 million years ago)

China (before 1 million years ago)

It bears repeating that, in their daily life, *erectus* communities had to care for children and strategise together. They needed to plan things like what to do today, where to hunt, or which men stay with the women and children and which go to find food. They needed to share information about signs ahead, about evidence of animals in the vicinity, or how to care for their sick, even if that amounted to little more than feeding them. It is, of course, speculation to imagine how they did these things or how well *erectus* communities understood or planned what they were doing, how they cared for each other, or how they conducted their daily lives. But by using the examples of current hunter-gatherer populations, along with the intelligence that *erectus* needed to have based on archaeological evidence, these suggestions are probably not too far off.

Erectus communities also had to learn to evaluate others and deal with them. There would be cheaters and laggards on the journey. Perhaps murderers. There would have been injured people. They would have desperately needed to work together. These pressures developed their intelligence and cultural connection more each day, along with their values and priorities.

Erectus did not simply walk single-file or run randomly around the world. They were organised. They were smart. They were a society of cultured humans. And they must have had language.

And yet what is language after all? With all this discussion of the language of *Homo* species, it is time to examine the nature of human language in more detail.

4

Everyone Speaks Languages of Signs

... by 'semiosis' I mean ... an action, or influence, which is, or involves, a cooperation of three subjects, such as a sign, its object and its interpretant ...

Charles Sanders Peirce (1907)

WHAT IS LANGUAGE? Is language indeed something that *Homo erectus* invented? It is worth restating the basic principle: language arises from the convergence of human invention, history, physical and cognitive evolution. The inventions that would have moved humans towards the languages spoken today were first icons and then symbols.

The archaeological evidence in fact supports the order predicted by the sign progression of C. S. Peirce – indexes would have come first, followed by icons and then symbols. We find indexes earlier than icons and icons earlier than symbols in the prehistoric record. Moreover, indexes are used by perhaps all creatures, icons recognised by fewer creatures and symbols used habitually only by humans. Although Peirce in fact believed that icons were simpler than indexes, he primarily had in mind the human elaborations to indexes, not – in my opinion – how the signs are found in nature per se.*

Newspaper headlines, store regulations, movie titles and other unusual forms of modern language are occasionally reminders of how simple language can be. There are some famous examples of languages reminiscent of possible early languages in the movies:

---

*If Peircean scholars disagree with my interpretation, then this is where I must diverge slightly from Peirce.

You Jane. Me Tarzan.
Eat. Drink. Man. Woman.

And in store signs:

No shirt. No shoes. No service.
No ticket. No wash.

These examples can even be found on billboards: *You drink. You drive. You go to jail.*

In spite of their grammatical simplicity, we understand these examples just fine. In fact, one can construct similar sentences in any language that will be intelligible to all native speakers of the language, as in these examples from Brazilian Portuguese.

Olimpiadas Rio. Crime, sujeira.
Olympics Rio. Crime, dirt.
Voce feio. Eu bonito.
You ugly. Me pretty.
Sem lenço. Sem documento.
Without handkerchief. Without document.*

Such phrases are interesting because they prove that humans can interpret language even when it isn't structured grammatically. *Homo erectus*'s language might have been no more complicated than these examples, though quite possibly it was more intricate. What all of these examples show, however, and what would have also held for the language of *Homo erectus*, as for all of the languages of *Homo sapiens*, is that language works fine when it is underdetermined. In understanding language or people or cultures, context is crucial. It is necessary to take a holistic perspective on interpretation. What was the organism,

---

*This last phrase is actually a slang expression that means that someone left in haste. During the time of the military dictatorship in Brazil, from 1964–1984, it was also a daring phrase because if caught without a 'document' (shorthand for the national identification card), a citizen could be arrested, and even tortured if they had the wrong political background.

its connection to the environment, and the thing it invented like? These are the questions that flow from a holistic perspective on the invention and evolution of language.

This idea is explored in detail by anthropologist Agustin Fuentes of the University of Notre Dame in Indiana. He makes a case for an 'extended evolutionary synthesis'. What Fuentes means by this is that researchers should not talk about the evolution of individual traits of species, such as human language, but instead that they need to understand the evolution of entire creatures, their behaviours, physiology and psychology, their niches, as well as their interaction with other species. Fuentes asserts that a full picture of human species engages the biological, the cultural and the psychological simultaneously as part of this extended evolutionary synthesis-based understanding. At the same time, Fuentes claims that current models of what culture is and how it interacts with the human psyche and body are poorly developed, at least in the sense that there is no broad consensus on what culture is. But there do seem to be components of culture and ways that such culture interacts with us. Many of the very traits and properties of the environment that we want to explain as part of language evolution are poorly defined, lacking widespread agreement on their meanings among the majority of specialists. For a theory of language evolution an understanding of the roles of society, culture and their interaction with individual cognitive functions is vital. Yet there is little agreement as to what any of these things mean. Although our bodies are a bit better understood, there are vast spaces of disagreement even about our physical make-up.

In order to better understand the factors of our environment that affect our evolution, it might help to start by defining the social environment, beginning with the elusive idea of 'culture'. A theory of culture underlies an understanding of language evolution. In fact, there can be no adequate theory of language evolution without a sound theory of culture. One idea of culture (mine) is the following:

> Culture is an abstract network shaping and connecting social roles, hierarchically structured knowledge domains and ranked values. Culture is dynamic, shifting, reinterpreted moment by moment. The roles, knowledge and values of culture are only found in the bodies (the brain is part of the body) and behaviors of its members.[1]

Culture is abstract because it cannot be touched, or seen, or smelled – it is not directly observable. However, the *products* of culture, such as art, libraries, political roles, food, literature, science, religion, style, architecture and tolerance or intolerance, are non-abstract, visible and tangible. Culture as a dynamic force is found only in the individuals of a society. Members of any society share a culture when they agree on a range of values and the relative priority of all the values that they hold. Members of a culture in turn share knowledge and social roles. One observes values and knowledge applied or examples of expectations from different social roles in action through individual members of the society. This is culture in action.

Each modern human, as did every *Homo erectus*, learns their place in society, what is more or less important as members of that society, as well as the knowledge common to all the members. And they teach these things by word and example to their offspring. All humans, past and present, learn these things. As do other creatures.

Nowadays, a very different theory of the origin of language than what I am urging here is popular among some. This is the idea that language is a disembodied object, along the lines of a mathematical formula. In this view, language is little more than a particular kind of grammar. If that kind of grammar, a hierarchical recursive grammar, is not found in a communication system, then that form of communication is not a language. Proponents of this idea also maintain that grammar 'popped' into being some 50–65,000 years ago via a mutation. This suggestion, even though very widely accepted, has surprisingly little evidence in its favour and turns out to be a poorer fit with the facts than the idea that language was invented, but subsequently changed gradually through all *Homo* species, to fit different cultures.

Though language is best understood as an invention, the mutation proposal is very influential. The theory comes from the work of Noam Chomsky, who began publishing in the late fifties and is, according to some, now the leading linguist in the world. But Chomsky's view that language is a recursive grammar, nothing more nor less, is a highly peculiar one. Already in 1972, a review in the *New York Review of Books* by American philosopher John Searle noted how strange Chomsky's conception of language actually is.[2]

This view is unusual because we know that languages need not have intricate grammatical structures. Some might instead merely juxtapose words and simple phrases, allowing context to guide their interpretation, as in the examples that begin this chapter. The main problem with the idea that language is grammar boils down to a lack of appreciation for the source and role of meaning in language. The view here, to the contrary, is that grammar is helpful in languages, but that different levels of complexity are to be expected among the languages of the world, including the extinct languages of *Homo erectus*. Moreover, complexity can vary tremendously from language to language. In other words, language is not merely a synonym for grammar. It is a combination of meaning, form, gestures and pitch. Grammar aids language. It is not itself language.

Language, whatever its biological basis, is shaped by psychology, history and culture. I try to show what this means in Figure 7.

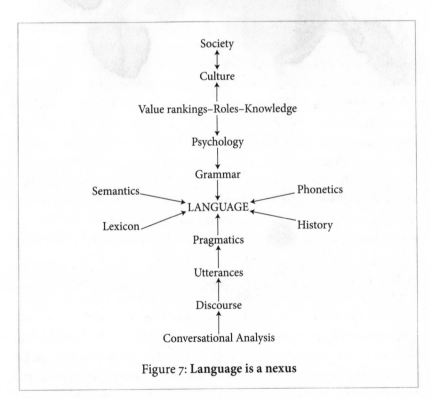

Figure 7: **Language is a nexus**

In order to get down to the nuts and bolts of how language itself actually evolved, there are two alternative views of development that must be distinguished. These are *uniformitarianism* vs *catastrophism*.

Uniformitarianism is the idea that the way things work now is the way they worked in the past. That is, the forces that operate in the world today are the same forces that have shaped the world since it began. Uniformitarianism does not deny the possibility of cataclysmic or catastrophic events playing roles in history and evolution. After all, uniformitarian scientists accept that there was a great dinosaur extinction event around 65 million years ago, when an asteroid crashed into the Yucatan. But it says that catastrophic change is not the main driver of evolutionary theory and that catastrophes should not be proposed as explanations without very clear evidence.

Catastrophism, on the other hand, appeals to major upheavals, such as Noah's flood or elevated rates of mutations, as frequent explanations for the origin and development of life on earth. Niles Eldredge and Stephen Jay Gould proposed that a great deal of evolutionary change is brought about by sudden macroevolutionary jumps that they called 'saltations'. Saltational models might be accurate for some examples of evolutionary change. But they always require additional evidence.

Uniformitarianism rather than catastrophism is taken to be a foundational truth in most scientific disciplines. In physics, few question the assumption of uniformitarianism. Physical laws show no evidence of having changed during the universe's natural history at least subsequent to the 'Big Bang'. And in geology, Charles Lyell's 1833 work, *Principles of Geology*, is known in part for its advocacy of uniformitarianism in earth history studies. By assuming uniformitarianism, a model of natural selection is expected to account for the transformation of ancient life forms into modern life forms via gradual, homeopathic, 'baby steps'.

In the case of language evolution, there are good reasons to reject catastrophism-based views such as Chomsky's. Reasons include its poor account of the genetics involved and its failure to account for the influence of culture on language emergence. Moreover, this catastrophism view fails to account for the fact that mutations for language are superfluous because language evolution can be explained without them. Invoking mutations without independent evidence is unhelpful.

In fact, the idea of language as a mutation simply offers no insights at all that help to understand the evolution of language. That is to say that language evolution can be explained without mutations, based instead on gradual, uniformitarianist assumptions, rendering superfluous proposals of language-specific genes or language-specific mutations.

Of course, one is free to propose mutations or anything else in order to construct a theory. And mutations are indeed among the drivers of evolution. But the rule of thumb in proposing mutations in the evolutionary record must be that 'in the absence of evidence, do not assume miracles'. And the proposal of a mutation as a crucial link in human language evolution must come with a full story of what evolutionary forces were at work at the time of the mutation that led to the spread of the mutation; otherwise, it is just invoking a miracle. Even if language ability were a mutation, it could only have produced an ability to learn language at a time when, ironically, there was no language. Apart from this lack of synchrony between need and mutation, someone proposing such a mutation must also explain what the survival advantage of a particular trait, such as language, was at the time of the mutation. And such an explanation must go beyond speculation to evidence. It could not merely be that 'language or grammar made thought clearer'. That is in all likelihood correct, but it doesn't speak to how or when language or grammar came into being. Nor does it offer any details on how it spread, either genetically or culturally. Otherwise, bandying about the word 'mutation' is unwarranted and speculative. This is the major weakness of the saltationist or what one might refer to as the 'X-Men' theory of language origin. Moreover, such a conjecture is unnecessary. Good old-fashioned Darwinian natural selection offers a more scientifically grounded story.

The sudden-emergence theory of language origin, echoed by palaeontologist Ian Tattersall in several works, also depends heavily on arguments based on absence of evidence. In this case, imagine a picture of a hawk in the sky accompanied by the caption 'There are hawks here.' Barring photo-altering software, this picture is pretty good evidence for the caption's veracity. On the other hand, a picture of a clear sky with the caption 'There are no hawks near where I live,' is much more problematic. The latter photo shows only an absence of evidence. It does not show solid evidence of the absence of hawks. It could be nothing more than a coincidence that the photograph failed to capture a hawk in a sky otherwise frequently populated by hawks. What is needed in such a case are more data, such as the lack of flora and fauna that hawks prefer along with the proper climate to sustain hawks. And identical reasoning applies to the unwarranted claims that language originated as a mutation, or that *Homo erectus* lacked symbolic representations. If they were right, of course, and *erectus* lacked symbols, then it would make sense to deny language to *erectus*. But in fact all that one is entitled to claim is that no one *had noticed* any evidence for such representations to date. But such evidence does exist in the record of the *erectus* cognitive explosion marked by their migration from Africa.

Moreover, there are languages spoken in the world today whose grammars have aspects reminiscent of what *Homo erectus* languages might have been like – namely symbols ordered according to cultural conventions. In this kind of case symbols follow an order agreed upon by members of a particular society. For example, Americans and Britons prefer to say 'red, white and blue' rather than 'white, red and blue' when discussing their national flags. Symbols and ordering can sometimes be vague and ambiguous, and therefore *erectus* would have needed the ability to use context and culture to interpret fully what others said.

To elaborate the nature of the absence of evidence argument, consider again what is known today about Amazonian tongues, which are unquestionably full human languages. But what record would exist of these languages if all their speakers died out and archaeologists discovered their speakers' bones 500,000 years hence? Forgetting for now that linguists and anthropologists have published grammars, dictionaries and other studies of Amazonian communication and culture, would these cultures and languages bequeath any material evidence that they were capable of language or symbolic reasoning thousands of years hence? Likely not. As stated earlier they would leave even less than what is found for *neanderthalensis* culture, aside from the few cultures that make ceramics, such as the famous Marajoara culture discovered on the Switzerland-sized delta at the mouth of the Amazon. It would be nearly impossible to find direct evidence that they had language – just as is the case for many ancient hunter-gatherer groups.

We are also unable to prove that the *Homo sapiens* who originally left Africa, or *neanderthalensis*, or Denisovans, or *erectus* had language, though it would be astounding if they did not, based on the cultural evidence. Therefore, one must be careful not to conclude that earlier hominins lacked language, merely based on the absence of evidence of artwork or what are commonly recognised as symbols in the prehistoric record. In the absence of evidence, the simplest idea about language evolution is that language gradually appeared via natural, incremental processes, following the invention of symbols, which in turn were made possible by the gradual evolution of the human brain and culture. This means that the burden of proof is on those who propose a sudden mutation for language, not on those who analyse the evolution of language as a gradual, uniformitarian process, fitting in with the rest of what we know about human evolution.

In recent years several palaeoanthropologists have inferred links between toolmaking and language evolution.[3] These researchers do not employ the 'absence of evidence' argument. That said, these studies appear to be based on an unusual conception of language as largely grammar and words, failing to consider the overall role and origins of symbols from abstract culture. They examine the growing complexity of tool use and relate this to a purported increase in language complexity, based on the assumption that the syntax of modern languages will always include complex syntactic devices for combining symbols such as hierarchy and recursion. In addition, these researchers discuss the absence of symbols among earlier *Homo* species, contrasting this with the widespread use of symbols among *sapiens*.

Efforts to explain language based on the archaeological record are admirable. Unfortunately, they often borrow bad ideas from linguistics to make their case. The principal bad idea is that quotidian objects are not themselves symbolic. Tools *are* symbols when they are the product of a culture. If one finds tools in conjunction with evidence for shared values and knowledge in a society, there is no need to look any further than the tools themselves for symbols. Tools may not be symbols for chimps, but they would have been symbols for *erectus*. Second, grammar does not require complex syntax and, therefore, neither does language. There are many groups today that have well-functioning, completely adequate languages but lack the kind of complex syntax that palaeoanthropologists sometimes assume in attempts to correlate complex tools with complex language. But such cultures also use complex tools, an unexpected state of affairs for theories proposing steady parallel growth of complexity of language and tools.

The understanding of natural language evolution must incorporate also the fact that language is a cultural tool for community building. More intricate syntactic structures of the kind found in many modern languages, such as subordinate clauses, complex noun phrases, word-compounding and others, are not crucial for language and are later additions made for cultural reasons. 'Expression of thought,' proposed by some as the raison d'être of language, is also a secondary feature of language. The evidence on hand from contemporary languages and the evolutionary record count against this in favour of communication as the primary function of language. At the same time, there is no doubt that the uses of language for thinking and communication are dependent upon one another. Each enhances the other.

If communication is the basic function of language, however, then human languages are not quite so unlike the communication of other creatures as some linguists, philosophers and neuroscientists assume. Communication is, after all, pervasive in the animal kingdom. Humans are simply the best communicators, not the only ones. But exactly in this quality lies the distinctiveness of human languages.

Alternative hypotheses should not, of course, be rejected out of hand, but only if the evidence is ultimately against them. The following quote is typical:

> [C]ommunication, a particular use of externalised language, cannot be the primary function of language, a defining property of the language faculty, suggesting that a traditional conception of language as an instrument of thought might be more appropriate. At a minimum, then, each language incorporates via its syntax computational procedures satisfying this basic property ... We take the property of structure dependence of grammatical rules to be central.[4]

Why do many researchers, such as those responsible for this quote, claim that communication is not the primary function of language? This idea flies in the face of what most people would regard as intuitive. Of course, the fact that a scientific idea doesn't match common sense, does not automatically make it wrong, since scientific judge-ment frequently differs from the average person's opinion. However, the reasoning here seems to be that humans are not all that good at communication. Take two well-known examples: 'Flying planes can be dangerous' and 'Visiting relatives can be a nuisance'. Both of these are ambiguous. The former example could mean 'Planes in flight can be dangerous because one might land on you' or it could mean 'Piloting a plane is occasionally risky'. Likewise for the latter example, which can mean 'Relatives who come to visit are not always welcome' or some-thing like 'A visit to my grandparents can prove tiresome'. For some, it follows that although ambiguity is found in communication, it is not found in thinking. Thus language works less well for exchanging ideas and information. If that were true it would then suggest that the principal purpose of language is to aid thinking, not communication.

This, however, doesn't follow. First of all, there are studies from the Department of Brain and Cognitive Sciences at MIT that explain why ambiguity is to be *expected* in a communication system.[5] It is produced by the need to keep the amount that must be memorised low while maintaining efficient communication. Therefore, if one says, 'I want two,' you know I mean 'two' and not 'too' or 'to' depending on the context. (This clearly is an English-only example, but homophones seem to occur in all languages.) Second, ambiguity and vagueness are rarely problems because context usually enables the hearer to pick out the meaning the speaker intended. If one says, 'He came into the room,' the pronoun 'he' is vague. One can only interpret 'he' if one shares enough information with their interlocutors to know who this particular 'he' refers to. Third, ambiguity in writing and speaking is not inherently problematic for language. Rather, ambiguity often is the result of bad planning. Thus, if someone began to cross a classroom and walked into student desks on the way to the door, the judgement would not be that this is evidence that walking is not for locomotion. Rather, one might conclude that walkers should watch where they are going. And the same goes for language – most ambiguity, vagueness and other shortcomings of speech can be avoided by planning and thinking before speaking or putting pen to paper. Planning for com-munication, like planning in most activities, is helpful.

There are other problems with the idea that language is not for communication. Evolution never designs perfect systems. Rather it builds jury-rigged devices piece by piece, using what it already has in place. Language, like everything else about natural life, is imperfect. Communication breaks down. But so does thinking! The assertion that one's thoughts are unambiguous to oneself is just that, an assertion. It needs to be tested. Another issue to note is that it is by no means clear that all people always or even most of the time think in language. Many people, such as biologist Frans de Waal and author Temple Grandlin, claim that they think in pictures, not in words. Experiments to check such purported patterns of thinking are called for.

In the earlier quote on language and communication, the authors claim that communication is a 'particular use' of 'externalised language'. This makes it sound rare indeed. They believe that the only kind of language that can actually be studied is so-called internal-language, or *I-language*. An I-Language is just what the speaker knows in order to produce their spoken external or *E-language*. French, English and Spanish are E-languages, but their speakers' knowledge of what underlies languages are their I-languages.

Although there are some who claim that we can only study I-languages, this is misleading. E-languages can also be studied. In fact, thinking about this more carefully, the only way to infer anything about a speaker's internal languages is to examine the utterances of their E-language. E-languages are the gateway to I-languages.

Moreover, our inferences about the analysis of any sentence or set of sentences are always based upon a particular *theory*. The interactions observed from speakers are the essential source of evidence of what speakers know, regardless of how we test them. Of course, it is an obvious fact that a given label for an E-language, say 'English', is an abstraction. After all, what is 'English'? British, Australians, Jamaicans and Californians all speak it in one form or another, but which is or is closest to 'real' English? How is 'real' English pronounced? What are its grammatical rules? There is simply too much variation in English around the world for anyone to say definitively what English is. Moreover, the sentences, stories and conversations that form the database for discussing English do not themselves exhaust the language. There are always data from some variety of English spoken somewhere in the world that have not yet been collected. It is in this sense, therefore, that

English is an abstraction. At the same time, the utterances one hears or the sentences one reads are not abstractions. They are the very concrete, empirical sources of what speakers know, what cultures produce and what people actually do. Asserting that one ignores what people actually say – their 'performance' as some call it – in order to understand their 'competence' (what people know about their language as opposed to what they do) is like claiming that college exams show nothing because they only measure performance (the answers students give), not competence (what students really know). Exams, however, exist precisely because performance is the only way to gauge competence. Whether the competence one wants to understand is knowledge of how to engage in dialogue, how to tell stories or how to produce individual sentences, one can only figure out what speakers know by what they do.

No one ever directly studies what people know. To assert that they do is a common error in thinking. Rather one infers knowledge through behaviour. It should also be remembered that the quote on page 74 ignores the fossil record, which makes it clear not only that language, culture and communication were part of the same cluster of socially evolved traits of human cognition but also that there was a slow semiotic progression fuelled by natural selection.

Moreover, seeing communication as the primary purpose of language facilitates the understanding of what is most interesting about language – its social applications. Thus, for many researchers, in the study of language grammar takes a back seat to things like conversational interactional patterns, discourse topic-tracking, metaphor, the usage-based accounts of grammatical forms, and cultural effects on words and how they are put together. Pursuing these ideas, and based on everything discussed so far about the evolution of *Homo* species, three hypotheses for the origin of human language come to the fore. Each of these takes a different view on the relative importance of when grammar arose in the evolution of human languages.

The first hypothesis is known as 'Grammar Came Last'. According to this idea, the most significant and first step in the evolution of language would be the development of symbols. Grammar is little more than an add-on. Language would have existed *before* grammar. In this idea, grammar required all of the rest of language to exist before it could become operational. In other words, language first needed symbols, utterances and conversations before it created a grammar to structure, and thereby enhance, our communication.

The second idea, a very popular one, is that Grammar Came First. According to this proposal, evolution of language is primarily about the origin of computational properties of language, such as syntax. Without these properties there is no language. Symbols, gestures and other components of language may have been around in some form previously, but patterns emerged that brought all of them together for the first time as a language. By this idea, there is no language without a very peculiar kind of computation. But a simpler idea is available, namely that the ability to 'chunk' words or symbols into ever-larger units – phrases, sentences, stories and conversations – is really the basis of all computation in language. This 'combinatoriality' informs our interpretation of words – without it in fact we cannot understand well the individual constituents of sentences. Think of a string of words such as 'if the girl is pretty then he will run up to her' and compare this string to 'run the up pretty her if then girl is will he to'. Structure guides the interpretation of the words and, over time, it more finely hones their meanings, bringing about nouns, verbs, prepositions and modifiers. Some take this checking to mean that form is a very specific kind of recursive, hierarchical entity. By this view language is more than what can be diagrammed along the lines shown in Figure 8 below.

This type of diagram is typically used by linguists to represent the constituent structures of sentences. Though it might appear complicated (it isn't really), the tree structure does indeed seem to be necessary in order to understand how modern speakers of English construct their sentences. In this example the graph represents one sentence, '... Bill saw Irving' as a constituent of the larger sentence beginning 'John said that ...' Likewise, the verb phrase '... saw Irving' is part of the larger chunk 'Bill saw Irving'. Furthermore, psychologists, cognitive scientists, linguists and others have demonstrated convincingly that such structures are not mere artefacts. They seem to reflect what native speakers of English know in some way about their language. The grammatical structures every native speaker knows are more complicated than any that they are taught in their composition classes.

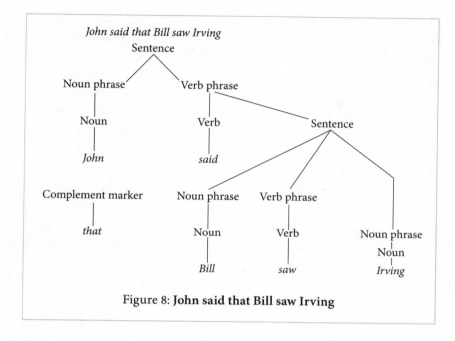

Figure 8: **John said that Bill saw Irving**

The third principal hypothesis falls between the first two. It is that Grammar Came Later. Though symbols came first, the evolution of language required a synergy between grammar, symbols and culture, each one affecting the other. In this view, structure, symbols and culture are co-dependent, together producing meaning, gestures, word structures and intonation to form each utterance of a language.

The notion that the form of sentences followed the invention of symbols might be interpreted in various ways. To understand this idea better, it can be instructive to view it in the context of the other two hypotheses.

Each of these possibilities accords a role to structure in language evolution. This is because linguistic form is of enormous importance in human communication and thinking. At the same time, claiming that the design of sentences is crucial for language raises questions. Perhaps the most important question is whether tree diagrams like Figure 8 are either necessary or sufficient for human language from an evolutionary perspective.

To proponents of the Grammar Came First hypothesis, hierarchical structure is the most important aspect of human language. And, again, many who adopt this hypothesis believe that language appeared suddenly, as recently as 50,000 years ago. According to them, not only did language not exist prior to *Homo sapiens*, but not even all *Homo sapiens* would have had language (since the species is more than 200,000 years old). A sudden mutation, say 50,000 years ago, would have not affected all *sapiens*, but only the descendants of the 'Prometheus' that won the language-gene mutation lottery. The idea that a mutation had an effect that ultimately left most of a species alive at the time at a disadvantage, or 'less fit' in Darwinian terms, is not unusual. DDT-resistance mutations have occurred in some insects, leaving them and their offspring the ability to thrive in a DDT-rich environment, while their conspecifics die off. But anyone proposing such a hypothesis obligates themselves to explain the evolutionary implications of their idea. The grammar mutation, to be of relevance to the evolution of language, must have made its possessor and his or her family more 'fit' than other *sapiens*, that is, more likely to survive. Or perhaps it was favoured by sexual selection, with better talkers becoming more appealing to the opposite sex and thus getting more sex and having more offspring. A third alternative is that a family with the language gene left the area and, via a population bottleneck, became the founder population for subsequent *Homo sapiens*, guaranteeing that all *sapiens* would have the mysterious 'language gene'.[*7]

This contrasts sharply with claims that language appeared very gradually over at least the past 3 million years and that all humans today have, and probably all *Homo* species in the past had, language.

Very importantly, hierarchical grammars, those whose structure requires tree diagrams, the kinds of grammars touted as crucial to language in many approaches to linguistic analysis and language

---

*If there could have been *Homo sapiens* without language 50,000 years ago, the hypothesised time of the 'Merge leap' for Berwick and Chomsky, that is more or less when the ability to do recursion entered human brains and languages, then there is no reason that there still could not be pockets of humans without language, that is without recursive thought or expression. This seems like a strange prediction, but it should be easy enough to verify. Finding humans that not only lack but who are completely unable to understand or produce recursive language would be striking support for the UG/recursion theory of language origins.

evolution, are simply by-products of information-processing tasks independent of grammar. Nobel-prize-winning scientist Herbert Simon wrote about this in the early 1960s in one of the most famous papers of the twentieth century, 'The Architecture of Complexity', Simon wrote that

> the central theme that runs through my remarks is that complexity frequently takes the form of hierarchy, and that hierarchic systems have some common properties that are independent of their specific content. Hierarchy, I shall argue, is one of the central structural schemes that the architect of complexity uses.

And also:

> I have already given an example of one kind of hierarchy that is frequently encountered in the social sciences: a formal organisation. Business firms, governments, universities all have a clearly visible parts-within-parts structure. But formal organisations are not the only, or even the most common, kind of social hierarchy. Almost all societies have elementary units called families, which may be grouped into villages or tribes, and these into larger groupings, and so on. If we make a chart of social interactions, of who talks to whom, the clusters of dense interaction in the chart will identify a rather well-defined hierarchic structure. The groupings in this structure may be defined operationally by some measure of frequency of interaction in this socio metric matrix.

Hierarchy would have been useful to *Homo* species as a way of understanding and constructing social relationships, of organising tasks and even structuring language. And we see such hierarchy in the organisation of the *Homo erectus* settlement of Gesher Benot Ya'aqov. But hierarchy is something that would only be needed in direct proportion to the growth of complexity in communication content – what is being talked about – as information flow grew faster and more complex. Information-rich communication, especially when coming at high rates of speed typical of human languages, will be aided, just as Simon predicted, by being structured in particular ways.

For example, the first three utterances will in all probability be harder for the average hearer to understand than the second three:

The moon is made of green cheese. Or so Peter says.
The moon is made of green cheese. Or so Peter says. Says John.
The moon is made of green cheese. Or so Peter says. Says John. Says
  Mary. Says Irving. Says Ralph.
Peter says that the moon is made of green cheese.
John says that Peter says that the moon is made of green cheese.
Ralph said that Irving said that Mary said that John said that Peter
  said that the moon is made of green cheese.

The reason is that the first set lacks recursion – sentences are all independent, side-by-side, but the second set does have recursion – one sentence inside another. It is a fact that, because of the complexity of the multiple quotes, recursion helps us to process the sentences more effectively. Although such sentences sound a bit artificial out of context, they are observed in English. Certain languages, however, those that lack recursion, can only produce sentences like those in the first set. As demands from the complexity of the society, such as the hearer or speaker interacting with more and more people they do not know, increase, so do the demands on grammar, though each society and language pair must be studied individually. It is possible to have complex grammars in simple societies, or the other way around.

An advocate for a mutation for language would have to explain why there are no *well-established* cortical specialisations for language or speech, aside from reusing parts of the brain for a variety of tasks, as many have claimed.[6] The growth of the prefrontal cortex, itself associated with toolmaking and sequential actions, helped to prepare the brain for language, by providing the cognitive firepower necessary for actions where procedures or improvised sequences are required. This is a form of exaptation, evolution's reuse of something that evolved for one task to perform another task, as in the use of the tongue, which evolved for food intake, in the articulation of speech sounds.

Grammars cannot exist without symbols. This entails that, even though grammars refine the meaning of symbols, grammars must follow symbols in the historical evolution of language.

Non-human creatures appear to use syntax. Therefore grammar is not exclusive to humans. Consider Alex the parrot, who, according to years of research by Irene Pepperberg, spoke (some) English and could understand even grammars with recursion and tree structures.

Humans have evolved away from cognitive rigidity (a by-product of instincts), to cognitive flexibility and learning based on local cultural and even environmental constraints. Under these assumptions, any grammatical similarities found across the world's languages would not be because grammar is innate. Rather they would indicate either functional pressures for effective communication that go beyond culture or simple efficiency in information transfer. An example of a functional pressure is the fact that in most languages, prepositions with less semantic content are shorter than prepositions with more content, as in the contrast between, say, 'to' or 'at' vs 'about' or 'beyond'. An example of efficiency in information transfer is seen in the fact that less frequent words are more predictable in their shape than those speakers use more frequently. So the verb 'bequeath', as a less-frequent verb, has a simple conjugation: 'I bequeath', 'you bequeath', 'she bequeaths', 'we bequeath' and 'everyone bequeaths' (this general principle is known as Zipf's Law). But the common verb 'to be', is irregular, as in 'I am', 'you are', 'he is', 'we are' and 'they are'. Both functional pressure and information transfer requirements optimise language for better communication.

Therefore, while grammar was neither first nor last in the evolution of human language, it necessarily came later than symbols. This conclusion is predicated on the evidence that in human interactions meaning is first and form second. Grammar does facilitate meaning transfer, but grammar is neither necessary nor sufficient for linguistic meanings.

Again, though, if grammar came later, what came first? Basically, two foundational advances were required of the genus *Homo* to start it on the road to language, both preceding grammar. We know this through the fossil record. Icons, indexes and symbols appear in the palaeontological record *before* evidence for grammar, just as the progression of signs would have it. Second, the prerequisite of culture is shown partially in the necessity of intentionality and conventionality in the appearance of symbols. Finally, languages without structure-dependent grammars

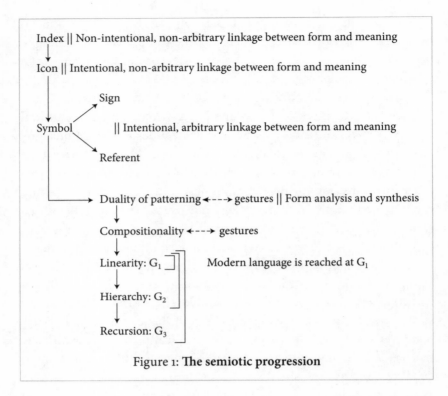

Figure 1: **The semiotic progression**

exist. The evolution of language followed instead the path of the 'semi-otic progression' shown in Figure 1 and repeated here.

Let's consider the components of Peircean semiotics presented in the diagram that are crucial to the evolution of language. First, there are *indexes*. Indexes are ancient, far predating humans. Every animal species uses indexes, which are physical connections to what they represent, such as smells, footprints, broken branches and scat. Indexes are non-arbitrary, largely non-intentional linkages between form and meaning. If an animal could not interpret indexes, then lions would never find prey, hyenas would search in vain for carrion and monkeys would be hard-pressed to avoid snakes and Accipitriformes (birds of prey). One can even *cultivate an ability* to detect and recognise indexes, as Native Americans, trained trackers, hunters and others do.*

---

*There is no language which has meaning without convention. Yet conventions imply culture (see my *Dark Matter of the Mind: The Culturally Articulated Unconscious*,

It is advisable to develop such an ability. On my walks through the Amazonian and Mexican jungles with different indigenous peoples, it was clear that they used indexes to know where they were, what flora and fauna were in their non-visible surroundings, where water was located and what direction would be best for hunting. People sniff, listen, look, feel and taste their way through the forest. Those unfamiliar with the indexes common to the jungle are often oblivious to the indexes they encounter, perceiving random smells, sights and so on, without recognising what they reference.

The deep knowledge of local index meanings can be referred to as 'emic' – or insider – knowledge.[8] Indexes are a vital rung in the ladder of human communicative evolution. And as they become enriched by culture, their significance in communication becomes even greater.

In a sense, indexes are a form of metonymical communication with nature, that is the use of the parts of something to perceive the whole (such as deer scat as a stand-in for the whole deer and the footprints of a horse for the horse itself). Even though the ability to recognise and interpret indexes can be culturally acquired, indexes are not enough to build a language. These indexes are inseparably connected physically to individual objects or creatures and therefore they lack arbitrariness and intentionality – two crucial components of symbolic language. Such primitive indexes have a limited role in human language because cultural rather than necessary linkage between form and meaning is essential to the latter.

Cultural linkage in the absence of a direct physical connection or resemblance between the meaning and the form associated with it dramatically increases the number of forms that may be used to link to meanings. In English we refer to a canine as *dog*. In Spanish as *perro*. And in Portuguese as *cão*. These are just arbitrary symbols for canines selected by these particular languages. There is no deep connection between the sounds or letters in 'dog' and the pet canine. It is simply what it is called in English. Thus its form is not necessary, culturally

University of Chicago Press, 2016) since they are general cultural agreements, such as word meanings. Finally, there is animal thought – if we say that language is structure-dependent and that it is a necessary condition for thought (as opposed to merely enhancing thought), then we are not only claiming that other hominins did not think, but that no other creature thinks, since other creatures lack the structure-building operation Chomsky proposes as the foundation of language, which he calls 'Merge'.

determined by convention. Lack of arbitrariness therefore means that indexes are unable to serve as the basis for language. But arbitrariness is a later step in the semiotic progression. It is preceded by intentionality. (Languages do have indexes where intentionality and arbitrariness have been added, going beyond the most primitive indexes shared by most species. These are words like 'I', 'here', 'this' and so on.)

Intentionality is the property of being mentally directed towards or about something and requires a mental operation or 'stance' of attention directed towards something. Intentionality is a property of all animal minds. Iconic signs, paintings, pebbles that look like faces, bones that resemble penises and so on, entail intentionality in representations of meanings because they are interpreted (and usually designed) to look, sound, taste, feel, or characterise the things they represent. These icons, such as the Makapansgat pebble or the Erfoud manuport or the Venus of Berekhat Ram, show some of the earliest steps from non-intentional indexes to the intentional creation of signs. The object is seen through a physical resemblance. The icon is 'about' something even if it was not intentionally created. Even non-iconic art, such as the geometric shell carvings from Java on page 95, shows the merger of intentionality and representation that is vital to all human languages. As intentionality meets representation in icons, humans were able, in principle at least, to begin to communicate more effectively. After all, modern day emojis are able to act as a kind of icon-based language. However, emojis depend on the modern grammars from which they emerge for complexity of interpretation and their organisation.

The next step is the most important of all the signs for language – the symbol. By being both intentional *and* arbitrary, the symbol represents a much longer stride towards modern language than either the index or the icon (although both are still found in all languages). Though symbols are often thought of as words, they can also be entire sentences.

Because of this evolutionary progression, which is gradual and spread across *Homo* species, no one woke up one day talking a modern, complex language, just as no ape woke up one day as a man or a woman. Nor, I believe, did anyone instantly acquire the ability to do recursive grammar-building operations in their head, in search of words to supply to such an operation.

Icons can shape languages in ways beyond mere images or

onomatopoeia, however. There are areas in which iconic sound representations are non-arbitrary, culturally significant components of human languages. These suggest that icons played a role in the transition to symbols that was so crucial to the invention of language. An example from the language spoken by the isolated Pirahã people indigenous to Amazonas in Brazil, which involves differences between men's speech and women's speech, helps to illustrate this.

First, Pirahã women use a more impressionistically 'guttural' speech than men. This is produced by two culturally motivated uses of the Pirahãs' vocal apparatus. One is that most Pirahã women's sounds are articulated further back in the mouth, relative to men's speech. Where a man might produce an /n/ by placing the tongue just behind the upper teeth, in women's pronunciation, an /n/ places the tongue further back, to just before the alveolar ridge at the roof of the mouth (which can be felt by running the tongue just behind the upper teeth).

Also, Pirahã women have one sound fewer than the men. Whereas Pirahã men have the consonants /p/, /t/, /h/, /s/, /b/, /g/ and /ʔ/ (glottal stop), the women use /h/ instead of /s/. For men the word for manioc meal is **ʔágaísi** whereas for women it is **ʔágaíhi**. This use of different articulation along with a different number of phonemes is a way of representing iconically via sounds the social status and gender of speakers. University of Chicago anthropologist Michael Silverstein has studied these kinds of language phenomena extensively, referring to them as 'indexical' markers of social relationships.

When using sounds, indexicals are part of a larger phenomenon known as *sound symbolism*, which has also been studied for quite some time. The Max Planck Institute of Nijmegen, in the Netherlands, has dedicated an entire centre to it.

The relevance for language evolution of sound symbolism, indexicals, Pirahã men's vs women's speech contrasts and so on is found in at least two potential stages of that evolution. The first stage is imitating sounds. This imitation can create words. Second, the use of sound symbols can build cultural relations and understanding of nature. While conducting research among the Banawás, I recorded a couple of men imitating (that is, using sound symbolism) the sounds of the animals that they hunted. Weeks later I played this tape for the Pirahãs, an unrelated hunter-gatherer group. Their response was 'Banawás know the jungle. They do these right.' This sound symbolism can be cultivated (as can the recognition of all indexes of the animals and other components of the ecological niche in which a group is found). The playing of the Banawás' sound symbolism for the Pirahãs was sufficient for the Pirahãs to know that the Banawás are more like them than Americans (whom they have never heard imitate an animal of their environment with accuracy).

According to the theory of Peirce, however, indexes, icons and symbols are still insufficient for language to get off the ground. One needs something in addition, which Peirce referred to as the 'interpretant'. This is in essence what makes it possible to use a sign so as to understand its object. Interpretants depend on certain aspects of a sign to work. As an example, consider the symbol 'eye'. This written word-symbol has two components – the letters 'e' and 'y' – that are separately usable, but here combined into a single English word to refer to our visible organ of vision. Written words are constrained by cultural conventions on the shape and order of the letters, e, y and e, that compose them. For this reason, if one rotates an /e/ to give its inverted mirror image, /ə/, then the interpretant of the letter, and ultimately the word it is part of, is lost. But write either a tiny /e/ or a ten-foot tall /e/ and nothing of the interpretant is lost. Thus size is not part of the interpretant of 'eye', though the directional orientation of the letters is. From this, it is seen that the symbol is itself analysed into meaningful parts that produce the interpretant.* Peirce was right again.

Of course, language evolution is also about biology, not only semiotics or culture. It is biology that underlies human language abilities. Acknowledging this obvious fact, it is perhaps surprising and counterintuitive to some to discover that there is nothing in the body dedicated to language. Not a single organ. Nothing in the brain. And nothing in the mouth (except for the position of the tongue). But this should not be a shock. Evolution always prefers tinkering with or exploiting what already exists over creating the brand new. What underlies our

---

*Like many of Peirce's terms, 'interpretant' has many different potential meanings, including the concepts of how the term would be translated and how one would interpret it. I refer here to one small aspect of this complicated network of meanings that Peirce intended by this term. It means much more than what I discuss here.

wonderful human voices is a jury-rigged collection of anatomical parts that we need for other things. This tells us that language is not a biological object but a semiotic one. It did not originate from a gene but from culture.

Every part of the vocal apparatus has a non-speech-related function that is more basic from an evolutionary perspective and that is found in other species of primates. Language and speech came later and exploited human bodies and brains as evolution had produced them, altering them over time. Therefore, it is not surprising that mechanisms implicated in human language, like our tongues, teeth and the rest, are not only part of the endowment of modern human biology but also found in other animals. This is a simple consequence of the continuity of evolution by natural selection of uniformitarianism. The single unique aspect of the human vocal apparatus that does seem to have evolved specifically for human speech, again, is its *shape*, caused by the position and form of the tongue (to which we return in detail later).

Sign languages also have much to teach us about our neural cognitive-cerebral platform. Native users of sign languages can communicate as quickly and effectively as speakers using the vocal apparatus. This means that our brain development cannot be exclusively connected to speech sounds, or else all other modalities or channels of speech would be unavailable or less good for language. It seems unlikely that every human being comes equipped by evolution with separate neuronal networks, one for sign languages and another for spoken languages. It is more parsimonious to assume instead that human brains are equipped to process signals of different modalities and that the hands and mouth provide the most efficient modes of physically expressing language. Sign languages, like spoken/oral-aural languages, also show evidence for syllable-like groupings of gestures. This means that humans are predisposed to such groupings, in the sense that our minds quickly latch on to syllabic groupings as ways of more efficient processing. This turns out to be very important in an understanding of the nuts and bolts of how language operates and how its operations were invented. Regardless of other modalities, though, the fact remains that vocal speech is the channel exclusively used by the vast majority of people. And this is interesting, because in this fact we do see evidence that evolution has altered human physiology for speech. Although humans can produce a rich array of sounds, they don't actually need to do so. By the use of only a small range of consonants (even just one), intermixed with one or more vowels, all human meaning can be communicated (and in fact could be communicated with only one vowel). We know this because there are modern languages that use very little of the sound distinctions provided by the modern vocal apparatus.

There is a long history to icons in the history of our species. After indexes such as fossilised footprints, icons such as the Makapansgat manuport are the oldest signs we have evidence for – *exactly as Peirce would have predicted.* These existed before symbols, following the predictions of the semiotic progression. For more than 3 million years visual icons have been collected by hominins, from *Australopithecus* to *Homo sapiens.* These icons suggest that the icon-possessor(s) quite possibly grasped a connection between form and meaning – what the icon is a visual representation of. In this light, consider the two-by-three-inch stone found in the Makapansgat cave in South Africa (Figure 9).

This pebble is much older than *Homo,* however. It was collected by none other than *Australopithecus africanus.* The manuport ('carried by hand') stands out among the tools it was found among because it clearly is not a tool, but was brought to the cave from elsewhere, almost certainly because it resembled a human face. And it is a kind of stone different from that of the cave where it was found. This manuport indicates that as early as 3 million years ago early hominins recognised iconic properties in objects around them. Just as one perceives the serpentine iconic properties of tree roots in the Amazon, so the australopithecines of Makapansgat saw iconicity in a rock with two circular indentations above a groove running transverse to them.

Someone might suggest that this manuport's human-appearing face was not noticed by the original australopithecine collector, but that only modern humans looking at it with our larger brains and language recognise it as symbolic of a human face. This proposition, however, would make for a much more complicated problem. We have a pebble among stone tools in a cave that was occupied by *Australopithecus africanus,* but although the tools were made there, the pebble wasn't. Now,

Figure 9: **Makapansgat manuport/pebble/cobble**

either it was *intentionally* carried to the cave or it was not. If it was, then why? One explanation is the one given – it was carried there because it looked like a face. The other explanation would be that it was wanted for something else. But its appearance looks like the simplest explanation. The proposition that it was taken there unintentionally, perhaps stuck in the mud between the toes of a returning creature, seems much less likely. So, although it cannot be proven that the pebble was brought back because it looked like a face, that certainly seems to be the best explanation available.

Move forward to 300,000 years ago and another manuport appears, this one in what is modern-day Morocco, picked up by *Homo erectus*, a cuttlefish bone shaped like a phallus (Figure 10).

Again, the icon was intentionally appropriated, recognised and collected, though not intentionally created. 'So what?' one might ask. What is the significance of such iconic objects for the development of language? Well, the answer depends on whom you ask. Chomsky's view of language as the output of a recursive operation, Merge, rules out any significant relationship between such occasional icons and the development of language proper, which is merely a type of grammar. If language is nothing more than a computational system, a set of structures embellished by local words, then clearly a phallic cuttlefish bone fails to move humans any closer to such a system. On the other hand,

Figure 10: **Erfoud manuport**

if language is about meaning and symbols, in which computation is nothing more than an aid to communication, then icons become vital to the reconstruction of the evolution of language.

Art, tools and symbols therefore each contribute to our understanding of the other and to the 'dark matter' of culture and psychology that allows each to emerge. Art is a visual form with shared meaning, the communication of emotions, of cultural moments, of ideas and so on via shared tacit cultural knowledge. It is necessary to learn to see in several different ways to appreciate art. If the art is painting, people must learn to recognise two-dimensional images of three-dimensional objects. If the art is sculpture they must learn to see in objects that are iconic or objects that are not quite iconic the real-world or imaginary object that the artist intended.

Tools, especially when they are generalised and found in different places, such as the 'kits' of early humans, indicate the existence of shared objectives, problems and solutions. For example the 300–400,000-year-old Schöningen spears (Figure 11) are evidence of culture among *Homo heidelbergensis*, perhaps a form of *Homo erectus*, and show that these humans hunted, that they used brute force rather than throwing and that they dedicated planning to hunting. Thus the spears represent cultural objectives, cultural knowledge, and cultural techniques. To members of the cultures that use them, they are therefore symbolic of

Figure 11: **A Schöningen spear**

these things, especially in light of the wider body of evidence for *erectus* culture.

Because tools are symbols, they also manifest a property most theoreticians consider crucial for language known as 'displacement'. This term refers to the sense of meaning that occurs when the object or referent evoked is not present, such as a song your mother enjoyed that reminds you of her when your mother is no longer around. Tools also have intentionality – they were the result of the mental focus of their creators. As symbols, they have a cultural meaning too, one that represents an activity, displaced from the form and meaning of the tool. A spear means 'hunting', even when the spear is not actually being used for hunting. Lastly, tools can show an aspect of Peircean signs – producing the interpretant – in that only certain parts of the tools are meaningfully connected to their tasks. The blade of a knife is more central to its meaning than the colour or material out of which its handle is made. An axe can be hollow, or of various materials. What matters is the quality of the edge for cutting. Thus tools, when examined carefully, show cognitive thresholds of association to meanings that are not entirely arbitrary (since the tool is non-arbitrary to the degree that it has a limited number of design options to carry out the task it is designed to perform) but still sufficiently arbitrary to count as a symbol.

In old Western movies the cavalry scout can tell which indigenous community made an arrow found in a victim. 'That's a Comanche arrow, Friend.' This cultural identification is symbolic, arbitrary, conventional and intentional simultaneously. It is symbolic, in other words. Arbitrariness is found in style, the cultural specificity of their task, or cultural regulations on the use of the tools.

Consider now one of the famous Schöningen spears. For their original owners these would have elicited thoughts of, thus symbolised, hunting, of bravery, of caring for their families and of death. Some of these spears were for thrusting, not for throwing. The *erectus* (or if one prefers, Heidelbergensis) male had to have his testosterone pumping to use this on a woolly mammoth. Run up there and shove it in! An elephant-sticker. Such a spear is as pregnant with meaning and symbolism as any portrait painted on a French cave wall.

Another example of tools as symbols comes from *erectus* shell carvings on Java (Figure 12).[9] These shell carvings are striking not only because of their age but also because of their location and the *Homo erectus* artist that made them. Unlike Flores, Java could be reached on foot by the *erectus* shell-carver across the now-submerged marshy land of the Sunda Shelf.

The geometric designs on the Java shell could represent merely a pleasant decoration, shaped in part by perceptual constraints of the brain that perhaps favour geometrical design. Or these marks could be symbols whose meaning is lost for ever. They might even represent something intermediate between icons and symbols, precursors to representation of meaning. I suspect that the first guess is correct. Nevertheless, we know that the designer, a *Homo erectus* man or woman, picked up a shark's tooth and pressed very hard and deliberately to record these shapes. Notice that the lines are solid and continuous, without breaks. To make such marks, this ancestral human would have had to press hard enough to cut through both the (now decomposed

Figure 12: *Erectus* **shell etchings from Java**

and missing) brown outer layer of the shell into the hard white shell proper. He or she would have had to carve without stopping or the lines would have some visible breaks in them. There is intentionality in these marks.

Whatever these designs indicate, they are at least a manifestation of intentional activity, perhaps iconic, perhaps symbolic – maybe they represent the waves of a sea voyage. Fascinatingly, they were made some 540,000 years ago. The shell themselves were used as scraping and cutting tools, perhaps even as weapons, as they were by some Native Americans as late as the seventeenth and eighteenth centuries.

Thomas Morgan and his colleagues in a 2015 paper assert a strong link between tool development and the emergence of language:

Our results support the hypothesis that hominin reliance on stone tool-making generated selection for teaching and language and imply that (i) low-fidelity social transmission, such as imitation/

emulation, may have contributed to the ~700,000 year stasis of the
Oldowan technocomplex and (ii) teaching or proto-language may
have been pre-requisites for the appearance of Acheulean technology. This work supports a gradual evolution of language, with simple
symbolic communication preceding behavioural modernity by hundreds of thousands of years.[10]

This is a growing area of research, linking toolmaking to language
evolution via brain development. Thus the presence of tools in a society,
because they can be interpreted as symbols themselves, offers evidence
that the toolmakers had achieved a form of symbolic representation,
though it sometimes exaggerates the degree of closeness between tool
complexity and linguistic complexity.

In discussing tools relative to language, one also looks to the qualities that both language and tools illustrate of culture, shared intentions
and the ability to match form and function. This is the conceptual basis
of symbols. Olduwan tools are the earliest known. They were used from
roughly 2.6 million years ago. Such tools, whose uses included chop-
ping, scraping and pounding, were probably invented by *Homo habilis*
(if one accepts this name as a separate, non-*erectus* species within the
genus *Homo*), or possibly by australopithecines, but the tools at the
Olduvai Gorge were clearly transported and manufactured by *erectus*.
The Olduwan tool kit in Figure 13 shows stones crudely shaped to work
as weapons and tools, like a hammer or a hand-axe. They would have
not been precision instruments comparable to later tools, but they
represent a step forward in hominin technology, serving perhaps as
precursors to culture.

To produce an Olduwan tool, a 'core rock' is struck on its edge by a
round 'hammerstone'. The striking produces a sharp, thin flake, leaving
conchoidal fractures on the core rock, as seen in the image. The flakes
are often reworked for other purposes.

The manufacture of tools requires planning, imagination (having an
image of what the final tools should look like) and, at least eventually,
communication of some sort for instructing others in how to make tools.
The sequential operations call upon the prefrontal cortex and produce
cultural selectional pressure for more cortical horsepower, more smarts.
However, this pressure might have worked, the larger prefrontal cortex of

Figure 13: **Olduwan tool kit**

earlier *Homo* toolmakers, relative to australopithecines, may be a response to it. Not surprisingly, therefore, about 1.76 million years ago, roughly 300,000 years after the rise of *Homo erectus*, Olduwan tools were joined by other *erectus*-manufactured tools, in particular a new type called the Acheulean (Figure 14). Many people suggest that this long, mysterious period without innovation, longer by far than (yet still reminiscent of) the Dark Ages of Europe, is due to 'low fidelity social transmission'. In other words, because *erectus* lacked language. But this is not a necessary inference. Cultural conservatism is a powerful and common force. It is always easier to imitate than to innovate, especially if a culture discourages innovation, as is still common throughout the world.

If *erectus* indeed possessed language, then is there a problem for my theory of *erectus* linguistic achievements, namely that it took hundreds of thousands of years for *erectus* to develop the technological advance of Acheulean tools? It is possible that this is correct, that although *erectus* invented tools and symbols during the first cognitive, and in all likelihood linguistic, revolution, about 1.9 million years ago, it took an additional 600,000 years, roughly, of evolution followed by invention to achieve language. Acclaimed palaeontologist Ian Tattersall makes the same suggestion in several works. Nevertheless, such a pessimistic conclusion does not follow.

It is known that human cultures, even in the twenty-first century, are resistant to change. Imitation is favoured strongly above innovation when what is being imitated still works fine, as several anthropologists have claimed.[11] The lag might have resulted from a lack of cognitive or linguistic development. But it might also result from the nearly universal principle of 'satisficing', in other words, nature tends to be satisfied with 'good enough', not striving for the best.[12] Or religious conservatism. It is indeed a surprisingly long time. But this 'innovation gap' does not wear its explanation on its sleeve. And in light of all other evidence, it does not alter the hypothesis that *Homo erectus* invented language.

Regardless of a long delay, *erectus* did eventually improve on its Olduwan tool kit. Though Olduwan and Acheulean tools overlapped in their use by earlier hominins, Acheulean tools were more advanced. They were carried from Africa to Europe by *Homo erectus*, with Spain being their earliest European destination, about 900,000 years ago. Acheulean tools were not created exclusively by striking stone upon stone as were Olduwan tools. They also involved shaping after flaking with bone, antlers, wood and other tools, which provided more control for the toolmaker. Also, Acheulean toolmakers preferred to use the cores over the flakes as the primary tools. So they were an advance over, but also as a complement to, Olduwan tools.

Building on Acheulean technology, *erectus* added other innovative improvements to develop the more advanced Levallois technique (ca. 500,000 years ago). In the spread of all of these tools, however, we see communication, if not in explicit instruction or linguistically, then in the revelation of the tools themselves to other hominins, as they spread and their use and design became known and emulated.

The Levallois technique required fine work along the edges of a core, followed by a final blow that lifted the flake, presharpened by the earlier striking. These tools were often made of flint, a more workable material, and thus had finer edges, as seen in Figure 15.

The complexity and uniformity of Levalloisian tools leads some to argue that language is implicated in their manufacture, in order to account for the error-correction assumed to have been necessary. But speaking is not absolutely required. Learning is often a matter of

Figure 14: **Acheulean tools** (https://en.wikipedia.org/wiki/Acheulean#/media/
File:Biface_Cintegabelle_MHNT_PRE_2009.0.201.1_V2.jpg)

observation, followed by trial and error under a watchful eye, with
very little verbal communication required even in modern societies.
However, some form of advanced communication does seem to be
necessary for feedback, even in language-minimal training. Moreover, there is no doubt that making tools together and the correction of
flawed techniques by learners would have favoured language development for instruction. And this was occurring with the first hominins,
to actually produce intentionally iconic and geometrical art. The idea
that *erectus* was capable of some sort of sophisticated communication,
such as at least a G_1 language, is supported not only by their art and
tools but also by their travels. It stretches credulity to believe that they
travelled over land and sea and developed the settlement patterns they
did without symbolic communication.

Some researchers suggest that icons like the Makapansgat pebble
and the Erfoud manuport may have led to new neural pathways for
recognising that one thing may stand for another. The evidence seems
too sparse to justify this statement, though I do agree that new ways
of thinking can lead to new evolutionary pressures on the brain. These
can then enhance the ability to comprehend representations more
complex than mere indexes. More importantly, the iconic manuports
might have led to cultural change, far before cerebral change.

An interesting further bit of evidence in the evolution of the symbolic
comes from early art, such as the 250,000-year-old Venus of Berekhat

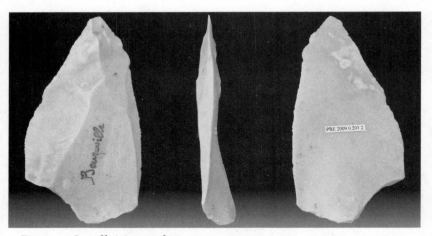

Figure 15: **Levalloisian tools** (https://en.wikipedia.org/wiki/Levallois_technique#/
media/File:Pointe_levallois_Beuzeville_MHNT_PRE.2009.0.203.2.fond.jpg)

Ram (Figure 16). Some deny that this is art, claiming it is nothing more
than a rock bearing a human resemblance, the same as the Makapansgat example. However, to some experts, it shows evidence, upon careful
examination, of having been manually altered to take on more of a
'Venus likeness'. And there is some suggestion of red ochre added to the
stone as a form of decoration. Though it may not have been a completely
ex-nihilo *objet d'art* in this sense, the evidence suggests strongly that this
is the oldest extant work of art in the world, either because it is carved
from scratch or because it is a human-modified natural formation.

The jump from index or icon to symbol is a relative baby step in conceptual development, though huge in language evolution. In my very
first encounter with the rain forest, I was constantly on the lookout
for snakes. Every tumescent root that 'slithered' across my path, partially covered in leaves, appeared to me first as a writhing, threatening
serpent and only secondarily as inert flora. Perhaps in recounting
similar experiences, two *erectus* buddies could have come to reinter-
pret the roots as icons for snakes, even eventually as snake symbols. (A
similar evolution is seen in comparing the earliest with the latest of the
Egyptian hieroglyphics or the Chinese writing system – icons become
symbols, that is, more arbitrary, with the passing of time.)

Symbols arise naturally within minds embedded in cultures, able
to learn, retain and integrate knowledge into a sense of personal and

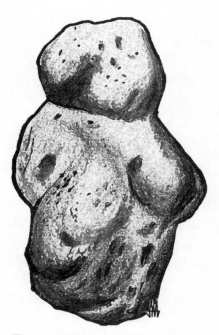

Figure 16: **Venus of Berekhat Ram**

group identity. One example, just given, is how the mind makes use of errors, perhaps moving from misperceptions to icons to symbols, one image 'standing for' another.

But they also arise from adaptation of the natural to the conventional in culture. One treatment of this route towards symbolisation is proposed by anthropologist Greg Urban. In his work on ritualised lamentation in Gê languages of Brazil, Urban argues that natural crying was transformed culturally into ritualised crying. This would illustrate a transformation of natural emotionally reactive sounds into a form of 'strategic vocal manipulation', a form of iconic representation of the emotional state of sadness. Further, he claims that 'strategic vocal deceptions in non-human primates are possible precursors of true socially constructed, socially shared metasignals, which in turn may be ancestors of modern human language'. Though newly created icons are insufficient for symbolic language, they do seem to offer a distinct and natural source of developing representations and thus perhaps a source of symbol invention.[13]

Another area in which symbols arise is in tracking social relationships. Most primates, among many other creatures, have elaborate social organising principles, through kinship, such as polyandry, polygyny, dominance relations, cross-cousins and parallel cousins. These concepts are learned via interactions, based initially on physical opposition, such as male vs female, strong vs weak, malleable vs non-malleable, or mother vs child. As people use concepts they come to understand them. So one can say, accurately, that even without language, many animals use something like concepts as they negotiate their ways through social relationships. Keeping track of such relationships would have increased the cultural and cognitive selectional pressures for symbols as some anthropologists in fact claim.

Numerous researchers have written on the evolution of the symbolic. However, as illuminating as these discussions are, they share a common lacuna, the connection of symbol-evolution and grammar to a well-developed theory of culture. It has been claimed that status symbols (such as expensive athletic shoes) have little to do with linguistic symbols. If correct, this would mean that a culture's use of status symbols is unrelated to whether they have symbols such as words. The most parsimonious interpretation of personal ornaments found in *erectus* burial sites is that these are nothing more than status markers. Part of the reason that researchers reject the potential linguistic significance of status symbols is because they claim that such symbols lack 'displacement' – a reference to something absent from the immediate context. Since we regularly talk about things that may or may not exist in other times or places, displacement is a fundamental feature of human language.

In other words, clothes and jewellery don't represent anything other than the taste and status of a person in the immediate context. But a minute's reflection reveals that this statement is incorrect. It lacks an adequate appreciation for culture. Status is inherent neither in ornaments nor in individuals, nor do ornaments bestow status. If someone finds the royal crown, putting it on their head not only does not make them a monarch, it makes them subject to the status-lowering charge of being an imposter. Status derives from culture. Status symbols are social signs. They are signs dependent for their meaning on abstract, displaced, cultural values. Thus although it is correct to say that status symbols are not linguistic symbols, both linguistic and status symbols are arbitrary, socially indexical and displaced. Therefore, they are conceptual kin. To have one is related to having the other. They would be expected to occur together in the same society at the relevant level of conceptual complexity or simplicity.

Displacement, the element which some claim to be missing in status symbols, is itself subject to cultural constraints. The crucial components for developing symbols are not displacement so much as arbitrariness and intentionality. But displacement is present for both status symbols and tools. Both kinds of artefacts reference abstract entities including the cultural values, social roles and structured knowledge that are present in the minds of all members of a culture.

What is the general evolutionary path to the development of symbols? To refer back to the example of roots across the jungle floor from earlier, when I see branches or roots at times as I am walking through the jungle, I jump back if I haven't had a good look, worried that one might be a snake. This mistaken association of one thing for another can lead later on to the intentional use of the object of the false impression to represent the thing it was falsely associated with. One could draw a root to mean a snake, or use the word for 'root' to mean 'snake'. As early hieroglyphic writing systems from different parts of the world show, this use of representations based on resemblance can further evolve so that all resemblance is gone, thus leading from an intentional use to an intentionally arbitrary symbol. As this happens, a drawing changes from an icon to a symbol. And just as this has happened in writing systems, so it is likely that it has happened in spoken systems, with different sound combinations becoming conventions, associated with a particular meaning.

In the research of evolutionary psychologists and anthropologists, one finds arguments to the effect that the development of kinship relationships would have created concepts in need of forms. That is, kinship relationships exert pressure on humans to go beyond icons to invent symbols. Concepts go looking for forms to serve as cultural exchange. I have a father. How should I communicate that to you? How should I say 'father'? But if, as many researchers believe, non-human animals too have concepts, then why don't animals develop symbols? One could retort that animals lack the language gene, but this is not terribly insightful, merely pushing the explanation back one level to the evolution of the gene rather than the evolution of symbols.

While there is no evidence for a specific language gene (the oft-cited FOXP2 gene is certainly not one, though it is sometimes claimed to be so), a great deal is known about the evolution of human intelligence and it is clear that humans are more intelligent than non-symbol-using creatures.[14] Thus a richer array of concepts requiring symbols and a richer, more inventive intelligence would have both been under pressure to find a joint solution to concept communication. Linguistic symbols arise to satisfy needs as cultures develop and they may emerge from status symbols, burial symbols and the like.

Anthropologist Michael Silverstein analyses the recursive properties of human thinking as applied to the use of language in representing cultural meaning, at multiple levels simultaneously. Another person exploring similar themes, explicitly linked to the recursive thinking (thinking about thinking or thoughts within thoughts) that underlies human cognition is Stephen C. Levinson.

Peirce anticipated both Levinson and Silverstein, however, in proposing that symbols are constructed of other symbols. In Peirce's writings, the phrase 'infinite semiosis' means that there is no limit to the number of symbols available to humans for languages. This in turn is based on the view that signs are multifunctional. Each sign determines an interpretant, but an interpretant is also a sign, so every sign embodies a second sign. This is a kind of conceptual recursion, concepts within concepts, and represents a huge step forward in human communication. It means that a string of signs always contains other signs. According to Peirce, this can be understood when we see infinity even in a simple sequence like:

$$\text{Sign}_1/\text{Interpretant}_1 \rightarrow \text{Sign}_2/\text{Interpretant}_2 \ldots \rightarrow \text{Sign}_n$$

This representation looks finite until we realise that Sign_n cannot be the end because if it lacks an interpretant it is not a sign. Likewise, Sign_1 cannot really be the beginning, because by definition it is connected to an interpretant of an earlier sign. So there is no beginning or end to symbols and signs. The process that creates them is infinite because it is recursive. Any random sign is always partially composed from another sign.

The origin and composition of symbols we have been discussing highlights the fact that, like other biological functions, human language is not simple. Language arises from interaction of meaning (semantics), conditions on how it is used (pragmatics), the physical properties of its inventory of sounds (phonetics), a grammar, phonology (its sound structure), morphology (the way the language creates words, such as by suffixes or prefixes or no additions at all) and the organisation of its stories and conversation. Yet, even after all this, there is something else. Language as a whole is greater than the sum of its parts. When we hear our native language we do not hear grammar or particular sounds or meanings, we hear and instantly understand what is being said as a whole, individually and together in a conversation or story.

Grammar not only is important for language to fulfil its culture-building function but also helps us to think more clearly. Yet in spite of the focus of many linguists on grammar as synonymous with language, grammar itself is no more important than any other component of language.

There are several reasons to reject the idea that grammar is central to language. First, languages like Pirahã and Riau (Indonesia) are languages currently spoken that appear to lack any hierarchical grammar. Their 'grammars' are little more than words arranged like beads on a string, rather than structured as chunks within chunks.[15] Second, there is a good deal of evidence that symbols evolved long before grammar in human linguistic history. Third, hierarchical grammars, when they are found, are little more than by-products. There are independent processing advantages of hierarchy well known in the computer science community, as Herbert Simon said. Hierarchical organisation aids tremendously in the processing and retrieval of any kind of information, not merely the information found in human languages. A fourth reason to reject the idea that any one type of structure is central to language is that non-human creatures appear to use syntax. But if this is true, ability to learn syntax is not exclusive to humans. Animals that use some form of linguistic structure include Alex the parrot and Koko the gorilla.* Their syntax is neither clearly hierarchical nor recursive,

---

*Koko is a female western lowland gorilla that has learned a good deal of American

but they nonetheless employ structure-based understanding. Fifth, humans have evolved away from cognitive rigidity. Animals need instincts because they lack flexible cognition. But this is the opposite of the direction that human evolution has taken, towards language and cognitive flexibility rather than instinct-driven behaviour as found among other animals. What humans know and learn is based on local cultural and even environmental constraints.[16] They are free to develop very different, non-genetically programmed structures. Similarities found across the world's languages would tell us about how human communication works, not about human evolution or hardwiring of human language instincts.

What did *Homo erectus* invent, then? Symbols. And symbols are just a short hop away from language. Over time the form and meaning units of *erectus* would have been ordered and perhaps structured, eventually producing ever more advanced structures, as in modern languages. But how did humans do this when other creatures did not? The answer is easy. All of human invention and language is underwritten, shaped and enhanced by the human brain. And, in fair play, language paid its debt to the brain by helping the brain become more intelligent, placing cultural and sexual selectional pressures on humans to communicate better.

The evidence thus strongly supports the claim that *Homo erectus* possessed language: evidence of culture – values, knowledge structures and social organisation; tool use and improvement (however slowly, compared to *Homo sapiens*); exploration of the land and sea, going beyond what could be seen to what could be imagined; and symbols – in the forms of decorations and tools. Only language is able to explain the *Homo erectus* cognitive revolution.

Language evolved relatively rapidly after the appearance of the first symbols. But as the benefits of hominin communication grew, so did the evolutionary pressures to produce clearer sounds, longer discourses

---

Sign Language. Her caregiver, Francine Patterson, claims that Koko can use as many as 1,000 signs accurately and that she can understand as many as 2,000 English words. Alex was an African grey parrot studied for over thirty years by animal psychologist Irene Pepperberg. Alex was claimed to have reasoning and linguistic abilities equal to dolphins and great apes. Pepperberg claimed that Alex could indeed understand the recursive G_3 language, English.

and more involved conversations. The story of the evolution of human language cannot be told fully without understanding how hominins evolved physiologically to support more complex and more efficient communication.

For this reason, we need to talk a bit about the evolution of our brains and vocal abilities.

Part Two

Human Biological Adaptations for Language

5

Humans Get a Better Brain

SCARECROW: Do you think if I went with you this Wizard would
give me some brains?
DOROTHY: I couldn't say. But even if he didn't, you'd be no worse
off than you are now.
SCARECROW: Yes, that's true.

The Wizard of Oz

IF SCARECROW WEREN'T a fictional character, he couldn't have been
simultaneously brainless and loquacious. Humans, of course, would
not be able to engage in conversations without a brain. But one doubts
that the poor, fictional Scarecrow understood what he was asking for.
If he could talk without a brain, he was probably better off. This is
because while brains are indeed the source of love, of sharing, of music
and of beauty, of science and art, they are also the origins of terrorism,
bigotry, war and machismo. The brain is simultaneously the reason for
our greatest accomplishments and the source of our greatest failures
as a species. But evolution doesn't care about success or failure in the
cultural sense and it certainly doesn't care directly about evil or beauty.
Evolution is about the physical survival of the fittest.

The hominin brain grew and developed for over 7 million years, from
Sahelanthropus tchadensis to *Homo sapiens*, about 200,000 years ago.
Then the growth and development seems to have stopped. There has
been no clear evidence for evolution in *Homo* brain size since *sapiens*
first left Africa. If *Homo sapiens* were smarter than *Homo erectus* and
Homo neanderthalensis 200 millennia ago, why are humans no smarter
today than those *sapiens* that first left Africa as the evidence suggests?
This could be due to any number of factors. It may be that there hasn't

been enough time for the brain to have evolved since *sapiens* first appeared; 200,000 years is a short period of time in the sweep of evolutionary history. On the other hand, according to some theories *Homo neanderthalensis* emerged from *Homo heidelbergensis* in only 100,000 years.

An alternative theory, the 'great leap forward' theory, suggests that change has occurred in the last 50,000 years, due to the appearance of art and leaps in cultural evolution. But there is no compelling reason to suppose that this change in the archaeological record is the result of biological evolution. Cultural development and new experiences could have built up slowly, finally leading to breakthroughs that would have seemed miraculous to earlier generations (such as the nineteenth-century Industrial Revolution). This is a time period long enough to have produced at least two or three such 'great leaps' in principle. So why is it that there seems to be no significant change in the brain for the past 200 millennia?

This apparent halt in human brain development is nothing to be ashamed of. It seems to be caused by the simple fact that life is good for our species. *Homo sapiens* have exploited a planet of plenty, through agriculture and technology, enjoying survival rates and quality of life that no other species has ever known. No other creature since time began has ridden the crest of the evolutionary wave as *sapiens* has, not even our human predecessors. *Erectus* and *neanderthalensis* never achieved cultural levels at which they could benefit from dentistry, science, relatively advanced medicine. They lacked the cultural resources to live with high mental and physical health and well-being. They lacked the intense innovation of *sapiens*. Was this the result of language? Did *Homo sapiens* have better language skills and therefore greater cultural accomplishments? The answer is, 'It's complicated.'

Language, as we have been seeing, is not that difficult, in spite of a long tradition going back to the 1950s telling us that it is extremely complicated, a veritable mystery. What we have seen, to the contrary, is that language is symbols and ordering at its core and that those are not tough ingredients to develop for brains like ours. On the other hand, having something to talk about can be hard. And that depends on both culture and individual intelligence. As *Homo* brains evolved and our intelligence as a species grew, it isn't so much that language improved but our ability to use it did. Smarter people can put the same tool to better use. And, yes, they can improve it. But the crucial thing here is the intelligence that our larger brains gave us over *erectus* to think even more abstractly and to take the symbolism our *erectus* ancestors had given us and project this into art and our stories. Into tool technology and so on. The combination of language and greater intelligence, accumulating knowledge over time, would have been all that was needed to result eventually in the second cognition revolution of a thousand centuries ago as *sapiens* emerged from Africa.

This greater intelligence, as well as growing relative geographical rootedness found in anatomically modern humans, such as Cro-Magnon – early *sapiens* in Europe – would have allowed them to construct more intricate cultures through greater social specialisation. Hunter-gatherer societies are often perfect examples of political anarchy, in the sense that they have no political structure other than group consensus. That has its attraction. Such societies often lack priests, full-time musicians, carpenters and all other specialised professions. This is because the cultural challenges accepted by hunter-gatherers (what they think is worthwhile, what their environment allows them to do, how they have chosen to live their lives) simply afford little opportunity for specialisation. Specialisation requires a society to provide food or goods to members who produce non-food related services or goods for the society. If someone spends all day playing a musical instrument to make someone else feel happy, they are going to need some food when they're done. But if no one gives them any, their music will stop while they plant a field instead of playing or singing. Language-enabled culture is the glue that holds the human cognitive colony together.

Once again, therefore, it is simplistic to suppose, as many researchers appear to, that the dramatically more complex cultural artefacts and social organisation of *Homo sapiens* relative to *erectus* or *neanderthalensis* are the result of language alone. *Sapiens* quite possibly have greater vocabularies and more complex grammars than other species of *Homo* had. The sapiens brain is better, but, more importantly, sapiens cultures and histories are richer. *Sapiens* have inherited much from other *Homo* species. They have incorporated ancient wisdom into sapiens cultures, languages and thinking. These accretions are in addition to all of the original developments, physical and cultural, by *sapiens* since they emerged from other species. Language has, of course, changed in the *Homo* genus over the past 1.9 million years or so. But a great deal of *Homo* biology has also changed. *Erectus* and others developed differently from the way in which we do.

Biological anthropologists have written on the different 'life histories' of *Homo* species. *Sapiens* develop more slowly than their *Homo* ancestors. Some of the life history distinctions between *Homo* species and other primates include longer pregnancies, longer periods of growth (sapiens infancy is longer, sapiens adolescence is longer, sapiens adulthood is longer than any other primate, including, apparently *neanderthalensis* and *erectus*). Humans have to live more slowly in order to live longer. This is common in the animal kingdom – slower growth usually means longer lives. Human biology confuses this simple picture a tad, however, because humans have very short periods between births, usually a feature of shorter-lived creatures. In this respect, humans are a cross between whales and rabbits.

If life history, brain growth and more nurturing as a result from parents, other kin and the culture at large also characterised part of the contrast between *sapiens* and, say, *erectus*, then these non-linguistic facts, already independently established in studies of biological anthropologists, could explain a great deal about the larger cultural and linguistic development of *sapiens* other than language. While there is nothing in the archaeological record to suggest that *erectus* lacked language, there is evidence that *erectus* were not as intelligent as *sapiens* and that they developed differently. But these points should not be confused. *Homo* species, including *erectus*, demonstrate several stages of brain evolution. By examining these stages, we get some idea of the advantages later species enjoyed over earlier ones.

Palaeoanthropologist Ralph Holloway and his colleagues proposed four major stages of hominin brain evolution based on years of research and study of the fossil record.[1]

Stage zero is the foundational stage, beginning with the split between chimpanzees and hominins. This stage extends back in time to *Sahelanthropus*, *Ardipithecus* and *Orrorin*, roughly 6–8 million years ago, when brains were marked by three characteristics that distinguish them from their descendants.

First, the lunate sulcus (crescent or moon-shaped groove in the brain) of these creatures is found further towards the front (anterior) portion of the brain. This groove divides the visual cortex from the frontal cortex. Since it is known that the frontal cortex of the brain is required for thinking, then, other things being equal, the larger this portion of the cortex, the better one can think. The position of the lunate sulcus is indicative of the relative thinking sophistication of the brain it is found in. Therefore, the further back the lunate sulcus is found, the reasoning goes, the more intelligent the animal.

Second, *Sahelanthropus* in all probability had a less-developed part of the brain dedicated to connecting multiple cerebral components. This portion is called the 'posterior association cortex'. This area of the brain links multiple regions simultaneously, enabling faster thinking. The posterior association cortex lets us bring several parts of our brains to bear on a single problem simultaneously.

Finally, the brains of the first hominins were small, 350–450cm^3 on average. This likely means that their smaller and more simply organised brains would have been incapable of anything like modern human thought.

The next phase, Holloway's stage one, of hominin brain evolution began about 3.5 million years ago, with the appearance of *Australopithecus africanus* and *afarensis*. The lunate sulcus in these creatures has moved a bit further back, relative to its position in earlier hominins. This is known because of impressions of the interior of their fossil skulls (endocasts). The visual cortex of the australopithecines had shrunk while their frontal cortex was now larger. Cognition was building steam.

The posterior association cortex is also larger in australopithecines. Their overall brains show signs of reorganisation and more specialised areas are becoming evident, along with an expansion in size to around 500cm^3.

Australopithecus cerebrums (the brain just under the cortex) show signs of asymmetry, with the left and right hemispheres taking on different specialisations. In modern humans this is very pronounced, leading to somewhat romanticised claims about 'left-brained' vs 'right-brained' personalities.

Hominin brains' next evolutionary jump occurs about 1.9 million years ago with the appearance of *Homo erectus*. By this time, the hominin brain had got much larger and specialised – an unparalleled combination of cognitive firepower. Nothing had been seen like this in the 4 billion years of evolution prior to the appearance of *Homo*.

This coincides with stage two in Holloway's scheme, marked by an overall increase in brain volume and encephalisation, accompanied by modern *sapiens*-like asymmetries (such as between left hemisphere for language and hearing from the right ear vs right hemisphere for hearing from the left ear and so on). At this stage, brains began to exhibit a prominent region around Broca's area* important for sequential actions. They also probably had better language abilities. There would have also have been increased development after birth of each *sapiens* and improved social learning in areas such as toolmaking, hunting and so on.

In stage three, Holloway's final stage that occurred about 500,000 years ago, the brain had reached its maximal size and refinement in specialisation for each hemisphere.

Therefore *Homo erectus* arrived on the scene with brain asymmetries typical of modern humans, such as a well-developed Broca's region. This implies the existence of, or at least the possibility for, some form of language. This is not a surprise, of course, since apart from directly looking at the brain of *erectus* there is evidence from their cultural accomplishments that they had language. These features of the early *Homo* brain mean too that *Homo* infants took longer to reach full maturity, since brain cells require the longest time to mature. It can therefore be inferred from these changes that *erectus* was capable of social learning in hunting, collecting, scavenging and reproductive strategies.

Now, it is important to avoid giving the wrong impression. *Erectus* was not the equal of *Homo sapiens*. In fact, compared to *sapiens* they had many, many shortcomings. It is important to discuss a few of the ways *erectus*, for all their relative brilliance, were inferior to *sapiens*.

First, their speech may not have carried over long distances. This is a result of their inability to form the same range of vowels that *sapiens* can produce, in all likelihood their vowels would have been hard to

---

*This is an area of the brain often identified with language. We will be discussing it in detail in chapters 6 and 7 below.

pick up across distances. On the other hand, like the Pirahãs and other groups, it is possible that *erectus* was able to overcome this shortcoming by simple yelling combined with distinctive pitch patterns. In either case, the fact that their speech would have not carried far doesn't mean that they did not have language.

Erectus speech perhaps sounded more garbled relative to that of *sapiens*, making it harder to hear the differences between words. This could have entailed less efficient communication than modern humans enjoy, but it would not mean that they lacked language. The existence of ambiguity, homonyms, confusion and the importance of context to interpret what someone has said to someone else continues to be crucial to modern speech. Part of the reason for *erectus*'s probably mushy speech is that they lacked a modern hyoid (Greek for 'U-shaped') bone, the small bone in the pharynx that anchors the larynx. The muscles that connect the hyoid to the larynx use their hyoid anchor to raise and lower the larynx and produce a wider variety of speech sounds. The hyoid bone of *erectus* was shaped more like the hyoid bones of the other great apes and had not yet taken on the shape of *sapiens*' and *neanderthalensis*' hyoids (these two being virtually identical). The non-modern *erectus* hyoid bone has profound implications for the evolution of speech and language, as will be seen.

These were not the only differences between *erectus* and other *Homo* species. *Erectus* faces were more distinguished by prognathism than modern humans', which would have impeded speech as we know it (though prognathism would not have blocked their speech).

Behind these physical differences were genetic differences between *erectus* and *sapiens*. The FOXP2 gene, though it is not a gene for language, has important consequences for human cognition and control of the muscles used in speech. This gene seems to have evolved in humans since the time of *erectus*. FOXP2 gives greater speech control. Possessing a more primitive FOXP2 gene, *erectus* would have had less laryngeal and therefore less emotional control in their speech. FOXP2 also elongates neurons and makes cognition faster and more effective. Without this *erectus* would certainly have been 'duller' than modern humans. But this is not a surprise.

Such a FOXP2 difference could have resulted in a lack of parallel processing of language by *erectus*, another reason they would have thought more slowly. FOXP2 in modern humans also increases length and synaptic plasticity of the basal ganglia, aiding motor learning and performance of complex tasks.

It is also unclear therefore whether *erectus* enjoyed the same degree of cognitive plasticity as we do. It is probable that *erectus* was a dull, non-inventive creature compared to modern humans. That doesn't mean that it was a languageless creature. As we have already seen, *erectus* was at the time the smartest entity ever to have lived. Just not as smart as *sapiens* turned out to be. The difference in intelligence might have been great or it might have been less than their brain size would indicate. There is much here that we do not know.

Supporting the idea of a less well-developed intellect for *erectus*, its most common tools were more similar in some respects to the tools of earlier, non-*Homo* primates. *Erectus*'s simplest tools may have been more homogeneous and non-combinatory (not built from multiple parts – handless axes vs axes with handles, for example). On the other hand, the earliest evidence for complex tools predates *sapiens*. These were hafted spears and were created by *Homo erectus* (or one of its descendants if a finer splitting of *Homo* species is preferred). And, of course, there are also the water craft that *erectus* used to cross signifi-cant distances in the ocean, which can only be classified as complex tools. Thus the archaeological record, while showing no complex *stone* tools, provides indirect evidence that *erectus* did in fact make complex tools of other materials.

To repeat, theories based on stone tools often omit evidence from non-stone tools. Palaeoanthropologist John Shea argues for a tight con-nection between technology and language, explaining that they are structured similarly in certain ways, though basing his work nearly exclusively on stone tools. This is understandable to a degree, of course, because stone tools are the only tools still available for direct study. And it may very well be true that if *erectus* had simpler technology then they would also have had a simpler language. This is not at all clear, however. Looking exclusively at stone tools is insufficient. It does not follow that simpler tools imply a lack of language or even a qualita-tively different kind of language. Some palaeoanthropologists appear to conflate complex toolmaking with complex syntax, by not being fully aware of the enormous variation among modern languages in this regard – some with complex etymological tools but syntax that is less complex perhaps than those tools would suggest.

Culture and biology together explain the apparent absence of ongoing extensive brain evolution among *Homo sapiens*. *Sapiens* seem to have passed a threshold of complexity that allows them to take care of themselves so well that they simply no longer have the same need for evolutionary assistance as they once did. As already discussed, this could come from different life histories in *sapiens*, cumulative cultural knowledge, language developed over time and differently powered brains. Modern humans live, survive and produce viable offspring because of culture.

This is not to say that there is no microevolution going on in modern humans. There could be humans with brains that are different from those of other humans in ways that produce greater numbers of viable offspring. But there is no evidence that brains are becoming larger or more specialised across *sapiens* either currently or since the beginning of this species. Neither is there a claim that sapiens brains could not evolve to someday make humans incomparably smarter than humans are today. One can imagine creatures with much higher intelligence on average than *Homo sapiens*. But evolution is not trying to build a brainiac. It is concerned merely with building a creature that is just good enough to have viable offspring.

And there is another thing. The only way natural selection can make people smarter is if more intelligent people have more offspring that live. But culture changes everything. Across the globe, cultures care for their members more effectively than at any time in human history. Cultural welfare has come to vie with physical evolutionary pressures in the definition of humans' evolutionary niche. Culture has also created a niche that is no longer purely biological, altering the course of evolution, as new cultural pressures arise and traditional biological pressures become comparatively less significant. Individuals who would have perhaps failed to survive without the level of cultural support available to modern humans are now able to transmit their genes to viable offspring. It may be that physically weaker or congenitally infirm individuals have no evolutionary disadvantage in the environment of a nurturing culture. This is good for humans, because cultural niches change, which favours increased diversity in the species, spawning ever more nurturing cultures, accelerating the change and survivability of those who at one time may not have survived. Eugenics advocated the improvement of human genetic heritage, but by failing to recognise the power of culture in shaping our evolution, eugenics had it wrong. Culture not only is the key to improving the species and the survivability of all, but also has liberated us from the strictly biological.

Humans arrived at this place of cortical stability through changes that might surprise some. They responded creatively and culturally to challenges of safety, travel, climate, shelter and food. As we learned earlier, they learned to cook food, which in turn helped them to eat more meat, which helped to shrink their guts. Repeating, then, calories that were once used for digestion were then freed up for *Homo*'s brains.

The result is modern brains and bodies, improved human thinking, morality and emotional control. This evolutionary progression reveals as clearly as anything the interrelatedness of organs and the embodiment of the brain in the body as a holistic apparatus. Human brains are smarter when our intestines are smaller. From *erectus* to *sapiens*, humans are in a sense self-made, pulling themselves along both the evolutionary and linguistic trails by their bootstraps. *Erectus* began the long process for humans of thinking their way to the modern world.

Looking back on the course of human brain and cultural evolution there are major discoveries, accelerated cultural evolution and long periods of stagnation among early humans.

Following the rise of *Homo sapiens*, profound innovations appear, along with a much faster rate of cultural change. This is why it is appropriate to refer to the *Homo sapiens* era, relative to that of other *Homo* species, as the 'Age of Innovation'. The innovation of *Homo sapiens*, greater than that of any other species, increased exponentially with the beginning of agricultural economies around 10,000 years ago on (arguably) opposite sides of the globe, in both Sumeria and Guatemala. Even before the rise of agriculture, however, innovations seemed to occur after species reached both cerebral and cultural thresholds. But an 'Age', whether of invention, imitation, or iron tools, does not characterise its entire population. In the Iron Age, people still used wooden tools, and in our present Age of Innovation, most *Homo sapiens* do not innovate in any significant way.

To learn about the brains of humans and how they underwrite

language from several sources, one must turn to neurology, palae-oneurology, archaeology, linguistics and anthropology. One must learn through clinical and neuroscientific studies of neurodiversity, from people with disorders such as specific language impairment (SLI), aphasia, or autistic spectrum disorder (ASD). And humans need to compare their brains with those of earlier great apes.

As *erectus* learned first, no brain is an island. Human brains are networked. First, their brains are networked in their bodies, connected evolutionarily and physiologically to other organs. But, equally importantly, their brains are networked to other brains. As philosopher Andy Clark has claimed for years, culture 'supersizes' our brains. A brain is an organ connected to other brain organs in the sea of culture. This is a point worth emphasising. In fact, one cannot understand the role of the brain in language and evolution without this conception. It is why caution must be exercised before accepting the popular but very misleading idea that the brain is a computer, an artefact very unlike an organ. Indeed, computers lack culture.

The question to ask rather is how do the anatomy, functioning and overall architecture of the brain help us to understand the role of the brain in the body, one organ among many? And how does culture help us to understand the brain as part of a social network of brains? And finally, the $64,000 question for our purposes: what does a brain have to be like for its owner to have language? The best conclusion is that the brain is a general-purpose organ evolved for fast and flexible thinking. It has to be prepared for anything. And for that very reason it is freer of instincts or any other form of prespecified knowledge than other species'.

Humans are fortunate that natural selection widened rather than narrowed their cognitive options. This freedom illumines our use and possession of natural language and other advanced cognitive abilities. However, when we lose any of that freedom, through different cognitive or speech disorders, the nature of our brains is revealed more clearly. This is why it is necessary to examine carefully breakdowns of the ability to engage in normal linguistic activity. Language deficits such as these might interfere with normal participation in a conversation, composing or understanding sentences, or being able to use the right words in the right context. Surprisingly, what emerges from such study is that there is little evidence that human brains have genetically specialised tissue for language. This perhaps startling assertion is supported by the fact that there is no convincing evidence to date that there are specifically heritable linguistic deficits. Language deficits are rooted in other physical or mental problems.

This may come as a surprise, although it would be perhaps more unexpected to learn that there *was* tissue or neuronal networks specialised for language, since language results from human neuroplasticity, which is in part the ability of neurons to change to better fit the needs of their containing organism. And there is, of course, also synaptic plasticity – the ability of connections (synapses) between neurons to alter as humans learn, grow, or suffer brain damage.

And not just humans. It has been discovered that if the third digit of an owl monkey is amputated (a gruesome experimentation that I hope will be halted), then changes happen in the brain of the monkey. The brain of the owl monkey has distinct areas for each digit. Following amputation, the area associated with the amputated digit will be overrun by other brain functions. In other words, the owl monkey's brain is flexible. Human brains are even more so. Brains don't let perfectly good neurons stand idle if they are needed for something else. Like Arnold Schwarzenegger in *The Terminator*, human brains rewire around damaged areas and repurpose undamaged areas that are no longer needed for their original functions.

Human brains also undergo a tremendous amount of synaptic change during a lifetime. Brains literally change – adding more connections, thus more white matter,* in response to learning – to adapt to new cultural environments or pathology, such as brain damage. Synaptic pruning and the establishment of novel synaptic connections in the brain are particularly robust features of human brain development before puberty, leading to the naming of this period of human development and learning as the 'critical period'. It isn't clear whether this

---

*White matter is named for the white (because fatty) material (technically, myelin sheaths) that surrounds the nerve fibres connecting parts of the brain used for higher cognitive functions. Timothy A. Keller and Marcel Adam Just, 'Altering Cortical Connectivity: Remediation-Induced Changes in the White Matter of Poor Readers', *Neuron* 64 (5), 2009: 624–631; doi:10.1016/j.neuron.2009.10.018

stage is as crucial in theories of cognition (such as language learning) as is sometimes claimed, but it is certainly an important segment of human cognitive development and neural plasticity.

As mentioned earlier, the brain is not a computer. It is important to underscore this again in the present context because it is a core belief for many linguists, cognitive scientists and computer scientists. The desire to see the brain as a machine goes all the way back to Galileo's analogy of the universe as a clock.

The appeal of this analogy is obvious, since both a computer and a brain handle information. But conceiving of a biological organ, whether a brain or a heart, as a computer, is an impediment to understanding either. To take one example, the brain does not appear to be organised into separate modules (or working units) for different functions in the way computers are. Additionally, the brain evolved without intervention. It is biological. A common reply to this is that it doesn't matter what the computer is made of, only what it does and how it does it. And yet the biological stuff from which the brain is made cannot live apart from its interaction with biological tissue and liquids that link it as part of a system with its vital non-computational functions (such as love). One could build a computer from human neurons but it still would not be a brain. Unlike a computer it *does* matter what a brain is built of and where it is housed. Now one might reply that a computer is also part of a network, plugged into a power grid and connected to other computers and so on. But neurology is ultimately not the same kind of thing as electronics. Computers lack biological functions, emotions and culture.

Another difference is that computers do nothing unless they are running a program. While brains do not literally have software, there are those who attribute something like software, a 'bioprogram', to account for language learning. But that metaphor has failed to produce answers to the kinds of questions and facts encountered in the story of human evolution. There is just no source of conceptual content inborn in all humans. Concepts are never inborn, they are learned.*

---

*Philosopher Robert Brandom has made this point in his own work, in books such as *Making it Explicit* (Cambridge, MA: Harvard University Press, 1998), in which he offers some convincing reasons to believe that we acquire concepts only by using them to draw inferences – that is one can say that humans have a concept only after

As Aristotle put it, paraphrased by Aquinas, 'Nihil est in intellectu quod non sit prius in sensu' – nothing is in the intellect that was not first in the senses.

On the other hand, perceptual abilities (seeing, hearing, feeling, tasting and certain emotions such as fear) do seem to be inborn. This kind of innate physical predisposition is drawn upon in language acquisition and cultural evolution. Some people use vision more than hearing when gathering data. Humans' emotional need for one another and desire for social interaction favour the development of language. So the brain definitely has specific, individual properties. But it is still important to avoid conceiving of the brain as a blastula of specific conceptual regions or a computer or pre-programmed with actual knowledge about anything.

One of the reasons that some creatures skipped the handout of brains from the evolutionary deity is energy consumption. Brains are calorifically expensive. The average human brain burns around 325–350 calories per day. That is about one-fourth of the average human's daily calorie consumption at rest (1,300) and about one-eighth of the average active person's requirement of around 2,400 calories per day. In other words, the brain is a high-maintenance piece of equipment. As specialists on the evolution of human fat consumption have observed:

> Compared to other primates and mammals of our size, humans allocate a much larger share of their daily energy budget to 'feed their brains'. The disproportionately large allocation of our energy budget to brain metabolism has important implications for our dietary needs. To accommodate the high energy demands of our large brains, humans consume diets that are of much higher quality (i.e., more dense in energy and fat) than those of our primate kin ... On average, we consume higher levels of dietary fat than other primates, and much higher levels of key long-chain polyunsaturated fatty acids (LC-PUFAs) that are critical to brain development.[2]

Beyond calorie consumption, another reason to lack a brain is

---

they have learned it well enough to use it in reasoning. I have made similar points from a very different perspective in Dark Matter of the Mind.

redundancy. Parasites can live in human guts without needing to think, ingesting whatever comes their way because of the brain-guided decisions of their hosts. They don't need brains because they use ours. Why waste resources? A final reason for lacking a brain is the absence of the correct evolutionary history. For humans this history was more complicated than for any other animal. Human brains, bodies and cultures have all evolved over the past 2 million years in a grand symbiosis. The body (including the brain) is connected to culture as hummingbirds are to flower pollination. Human bodies and brains are enhanced by culture, just as culture itself is enhanced by our thinking and language. Ever since Franz Boas, one of the founding figures of North American anthropology, it has been known that culture can affect body size, the use of language, what we recognise as 'talent', along with other aspects of human phenotypes. As noted in chapter 1, dual inheritance theory, also known as the Baldwin effect, refers to the discovery that culture indirectly affects the genotype itself. Natural selection favours changes in our alleles that produce culturally desirable components of our phenotypes.

To summarise what we have learned about brains, humans' larger brains could evolve only by overcoming three major disadvantages.[3] First, it has already been seen that brain tissue is among the most metabolically expensive in the human body. The second problem is that larger brains take longer to mature. Human children are not able to defend, feed, clothe and shelter themselves for a minimum of twelve years and sometimes much more, depending on the culture. Finally, the third major disadvantage of larger brains is that there is a conflict in bipeds between the benefits of narrow hips to aid movement and the need for a birth canal large enough to accommodate increasingly larger-brained infants. Big brains can kill mothers in childbirth because the birth canal is small. This is so that the mother can walk, while the brain is large, so that the child can think.

This raises the question of how big brains need to be to support human intelligence. Many palaeoneurologists use what is called the EQ or Encephalisation Quotient. This is a ratio of a species' brain size to the average brain size of a mammal with the same body size. The theory behind the EQ is that intelligence grows not so much with the absolute size of the brain (a sperm whale's is around 8,000cm^3), but with the ratio of the species' brain size to its body size. And this idea does seem to be a fairly reliable predictor. Tom Schoenemann from Indiana University has made the case that absolute brain size also matters, because it leads to specialisation in the brain that smaller brains are unable to achieve. Schoenemann lists several advantages to having larger brains for *Homo erectus*, other members of the genus *Homo*, and all other creatures.

First, 'highly encephalized species ... tend to forage (or hunt) strategically, taking into account the habits of their food (or prey), while less encephalized species tend to graze (or hunt) opportunistically'. Further, 'As brain size increases, different areas of the cortex become less directly connected with each other.' The consequence of this changed connectivity is that 'as brains increase in size, areas are increasingly able to carry out processing independent of other regions ... Such independence makes parallel processing increasingly possible and this has significant consequences because it leads to great sophistication in behavioral response.'[4]

Suzana Herculano-Houzel, in her 2016 book *The Human Advantage: A New Understanding of How Our Brain Became Remarkable*, has made the case that human brains are superior due in part to much greater neuronal density – we have more neurons per cubic centimetre overall and more connections between them.

The idea that culture affects behaviour, appearance, intelligence and other aspects of one's phenotype, leads to the conclusion that the most important question about our brains is not, 'What in the brain makes language possible?' The right question is, 'How do brains, cultures and their interactions work together to produce language?' The answer is that, over time, each has helped the other to improve. One cannot understand the evolution of language, therefore, without understanding the evolution of the brain. The brain likewise cannot be understood without understanding the evolution of culture.

The challenge in understanding the evolution of the hominin brain once hominins diverged from other primates some 6 million years ago (whether via *Ardipithecus*, *Sahelanthropus*, or *Orrorin*), is not only how the human brain got bigger, but why. It is known that the brain grew from the time of *Australopithecus* from approximately 500cm^3 to nearly 1,300cm^3 in a relatively short space of 125,000 generations, or 3 million years. To understand why this increase occurred, it is necessary to understand the brains of contemporary *Homo sapiens* and to work out methods for understanding brain evolution in light of fossil and cultural evidence. There were several changes in the environment that pressured human brains to enlarge in order to support greater intelligence. Fortunately, we know the starting point for these changes to be *Australopithecus*, and much is known about the end point of hominin evolution – *Homo sapiens*. It only remains to determine how we got from one to the other. This entails figuring out the stages along the way. It is necessary, therefore, to examine the evidence for brain evolution in the fossil record and changes in the environment that might have exerted selective pressure upon human brain evolution.

One aspect of the brain's growth and development, its encephalisation, is easy. Creatures with bigger bodies tend to have bigger brains. As fossils indicate an increase in the overall body size of hominins, they also show increases in brain size. The formula seems simple enough – grow the body, grow the brain. So did the brain just come along for the ride with the growth of the body? Maybe not. In fact, the relationship of encephalisation to body growth may have been the reverse. It is possible that the external pressures leading to the growth of the brain also caused hominins to have larger bodies. Brain size and body size are controlled by some of the same genes. As Mark Grabowski puts it,

> results suggest that strong selection to increase brain size alone played a large role in both brain and body size increases throughout human evolution and may have been solely responsible for the major increase in both traits that occurred during the transition [from *Australopithecus*] to *Homo erectus*. This switch in emphasis has major implications for adaptive hypotheses on the origins of our genus.

And, he continues,

> It may simply be that a larger brain requires a larger body to meet its increasing energetic demands and evolutionary constraints due to brain-body co-variation are one way of maintaining this relationship.[5]

This all means that the evolution of brain and body size is a chicken vs egg problem. Either the brain evolved and brought the body with it or vice-versa. But whichever came first, the enduring question is to understand the pressures that led to growth in human intelligence. I think that the best way to look at the problem is, like so many aspects of biological development and existence, as a case of symbiosis – where two or more creatures or parts of creatures (like brains) developed or evolved in tandem, each needing and affecting the other.

Considering, then, the implications of brain anatomy and functioning as part of a human body and culture for the understanding of brain and language evolution, it is worth returning to further discussion of palaeoneurology. What one needs to consider seriously is what it meant and still means for the understanding of hominin evolution that the human brain grew so quickly and reached such a large size relative to the rest of their bodies. As neurolinguist John Ingram has stated, this represents in evolutionary time a 'runaway growth of the brain'.

A fascinating discussion on brain size and growth is given by palaeoanthropologist Dean Falk, who compares the Taung baby fossil discovered by Raymond Dart to the discovery of the 'Hobbit', a tiny variety of *Homo erectus* whose fossils were unearthed on the island of Flores by Australian palaeontologists Peter Brown and Michael Morwood.[6] It had been known for some time that *Homo erectus* arrived on Flores and developed a robust cultural outpost some 900,000 years ago. But the Hobbit was unexpected. The first question about them was, 'Why were they so small?' Another was, 'How did they survive so long in co-existence with *Homo sapiens*?' Apparently, the Hobbit had lived until as recently as 18,000 years ago, perhaps even as recently as 14,000 years ago. Since most researchers believed that all non-*sapiens* species of the genus *Homo* aside from *neanderthalensis* had died out roughly 200,000 years ago, this was a shock.*

---

*David Gil, a researcher at the Max Planck Institute in Germany, has told me that in the folklore of contemporary local Indonesian communities, one hears of sightings of small, human-like creatures of the forest. It is fascinating to think either that the Hobbits might still exist or that stories about them might have entered the cultures of local Indonesian areas more than 18,000 years ago and are still told today. Less excitingly, though, it is more likely that the creature Gil has heard described is entirely fictitious, an invention of the local cultures. Its description's similarity to

The brain of these occupants, now known as *Homo floresiensis*, was much smaller than that of their *erectus* ancestors. In fact *floresiensis*'s brain was smaller than the brain size of many australopithecines, coming in at around 426cm³. What does this surprising reduction in brain size in the *erectus* line mean for the understanding of the development of human intelligence? Does the smaller brain size of the Hobbit creatures indicate that they lost intelligence? That would represent a fascinating step backwards in evolution. Both had similar brain sizes and both were around 3'11" (1.19m) tall, but was the Hobbit as smart as an australopithecine? Or was it smarter? Or was it not as smart? Was *floresiensis* as intelligent as any other *Homo erectus*, in spite of having a brain less than half the size of the *Homo erectus* that left Africa hundreds of thousands of years earlier?

Based on their use of tools and other archaeological evidence, it seems that *floresiensis* was smarter than *australopithecus*. There is evidence that it had culture, at least in the use and manufacture of tools, as well as its ancestors' initial voyage to Flores, discussed earlier. It is possible that the Hobbit had lost the culture of its ancestors, but this is unconvincing speculation, because we know it used fire and stone tools that were polished and shaped for working on softer materials such as wood and bone. This means, however, that intelligence is not and cannot be simply a function of brain size. There is no evidence other than cranial size to suppose that *Homo floresiensis* was any less intelligent than *Homo erectus*. If they were, in fact, equally smart then this raises the question of whether *erectus*, with its brain roughly two-thirds the size of a modern human's, might have even been as intelligent as *Homo sapiens*. The point of asking this question is that if one is looking for evidence concerning the intelligence of a fossil human, the cultural evidence may be more important than the physical. And the failure of brain size per se to reflect intelligence means that to understand the brains of our hominin predecessors we need accurate information about their cytoarchitecture, neural density, their culture and their languages. None of this is possible to achieve given current data and methods.

Summarising to this point, the archaeological record supports

----

small *Homo erectus* is probably coincidental.

the thesis that general intelligence underwrites language, not some hypothesised language-specific, innate portion of the brain. And no innate language-dedicated areas of the brain have been found. If this thesis is correct, then one might suppose that reliance on large-scale, non-innately specialised neurological connections ensures greater plasticity. The specialisation of regions of the brain is largely due to the cytoarchitecture of the relevant brain areas coupled with the ontological development of the individual (their life path), including their biology, culture and personal psychology. Yet overall, the brain draws on all of its forces largely simultaneously as its possessor moves through the world.

One lesson to be drawn, therefore, regarding the Hobbits of Flores is that inferring intelligence from brain endocasts is a risky business. There are certainly signs that can be interpreted, such as the evidence for development of different areas of the brain that we *know* to be associated with intelligence, language, planning and problem solving in modern humans, but the knowledge we derive from examining skulls is still inadequate for understanding the growth of human intelligence and must take a back seat to the cultural evidence. It would have been easy, in the absence of evidence about their villages, sailing, tools and so on, to claim that *erectus* was a dumb brute compared with modern humans because of its 950m³ brain size. But the cultural evidence, to the contrary, suggests that such speculation is unwarranted. What we see from the cultural evidence is that *Homo erectus* was intelligent, capable of human language and the master of its environment.

Robin Dunbar, a British anthropologist, claims that the main force driving hominins to develop greater intelligence was increased social complexity. Dunbar argues that it wasn't so much the problem solving required by ecological change that favoured the growth of human intelligence, but rather that the pressure for intelligence and encephalisation came from the increasing size of human societies. Humans were settling in larger and more complex groups. These surpassed in size and complexity those of any other primate. Dunbar's argument, then, has to do with the exponential growth in the number of social relationships that arises from even the modest increases in overall group size. Whereas humans' closest living relatives, chimpanzees, live in social groups of about fifty individuals, human hunter-gatherer societies average about 150, adding huge stress to the brain to keep track of the much greater number of social relationships that this 300 per cent larger group size entails. Individual members of a society are like neurons in the brain. The more there are, the more connections between them. In other words, just as it is the relationships between neurons that make the brain so complex, so it is the exponential growth in relationships as societal numbers increased arithmetically that required more intellectual horsepower in order to keep track of those relationships, at least according to Dunbar. In other words, as group sizes grew, the human cortex grew as well.

To support this hypothesis, Dunbar noted that cortex size co-varies with group size across several species. Of course, one could reply that Dunbar has the cart before the horse here. Perhaps it was brain growth and greater intelligence that enabled the growth in social relationships among humans, rather than the other way around? But the causality seems more likely to go Dunbar's way: social size → brain size, rather than brain size → social size. If one had a larger brain first, before social change, then one might have preferred to become a hermit. That is, the brain growing first could have led to any number of social models. But if society grew first then it would have indeed pressured the brain to be able to keep track of the new relationship sizes.

Another socially induced pressure for intelligence growth is the growth in cooperation. As humans banded together, they worked together. The first human bands were made viable through cooperative work. Of course, in any group effort there will usually be a passenger or two, those who are content to reap full benefits of the efforts of others while failing to provide full efforts themselves. For group relationships to work more effectively, therefore, natural selection would have favoured improved intelligence in order to detect cheaters.

As we have seen, sexual selection was noticed early on by Darwin as a major force in evolutionary change, responsible for beauty (such as male peacock feathers), physical attributes such as larger breasts in human females relative to other primates (apparently even early hominin males preferred their women to look pneumatic) and longer penises for human males.*

---

*Human males have the longest penises relative to body size in the primate

An additional consideration favouring increased intelligence could be that its possessors are more likely to survive diseases of the mind or nervous system (such as meningitis) that have the side effect of reducing intelligence in the survivors. This could have in turn fed sexual selection in that males and females would prefer mates who survived diseases with less impairment or long-term effects.

It is most likely that all of the above reasons contributed to the natural selective pressures for greater human intelligence. And yet none seems to be the primary contributor to major leaps in our cognitive powers. In fact, it doesn't seem prudent to suppose that 'smarts' can be understood exclusively through brain-case size of fossils or in overall size of the brain, or even in evidence that this or that area of the brain was better or less developed. Intelligence is not simply a function of brain size or brain component size. If it were, then among modern humans men would almost always be more intelligent than women because their brains are almost always larger, often much larger. Some modern European women have brains of only around 950cm^3, nearly identical in size to the brains of *Homo erectus*. Yet they certainly seem as intelligent as modern males with larger brains.

So what might have selected humans for greater intelligence as a function of brain size, cytoarchitecture, synaptic complexity, white matter, glial cells* and so on? The strongest force for the evolution of human intelligence was in all likelihood a combination of language and culture, as it is manifested through the use of symbols, grammar, pitch and gestures. As people began to use these methods of communication they were able to think more together, enhance each other's ability to know the world around them and to predict its future forms. Questions began to occupy the minds of sapiens ancestors: 'Where will that animal be in a few seconds?' 'In what direction will that fire burn?' 'When will the rain return?' 'To where does that river flow and what

---

world. This may result from the fact that humans are the only primates to engage habitually in face-to-face copulation. This in turn may have strengthened male–female bonding. Alternatively, and for whatever reason, human females may have been more attracted to better-endowed males.

*Glial cells and mast cells form part of the brain's neuroimmune system, independent of the immune system that protects the rest of the body.

will I find if I go up vs if I go down it?' And while asking these questions, humans needed to use language to bring order to their social interactions, naming kin and other relationships, leading to general improvement in their cognitive endowment.

Now that we have some idea of how the brain evolved in general terms, we need to ask the next question. What are the *specific* features of the human brain that underwrite our language ability? And are these features unique to language or do they serve other roles beyond language? That is the crux of a several-decades long debate in the cognitive sciences and palaeoanthropology.

6

How the Brain Makes Language Possible

The complexity of the nervous system is so great, its various association systems and cell masses so numerous, complex, and challenging, that understanding will forever lie beyond our most committed efforts.

Santiago Ramón y Cajal (1909)

HOWEVER THE HUMAN BRAIN reached its current state, *Homo sapiens* is now the proud owner of the greatest cognitive device in the history of earth. It is time, therefore, to ask how this device functions and how it is put together.[1] In particular, how does the human brain enable human language? Part of the answer to that question is that the human brain shares an organisational characteristic with the human vocal apparatus (the parts that help to create speech, including our lungs, tongue, teeth and nasal passages). Like the vocal apparatus the brain reuses pre-existing systems and exploits them for purposes other than what they might have originally evolved for – or at least what they were used for prior to their being marshalled for language use. This is a point made in the writings of numerous neuroscientists and philosophers. Neither the brain nor the vocal apparatus evolved exclusively for language. They have, however, undergone a great deal of microevolution to better support human language. It is often claimed that there are language-specific areas of the brain such as Wernicke's area or Broca's area. There are not. On the other hand, in spite of the lack of dedicated language regions of the brain, several researchers have shown the importance of the subcortical region known as the basal ganglia to language. Basal ganglia are a group of brain tissues that appear to function as a unit and are associated with a variety of general functions such as voluntary motor control, procedural learning (routines or habits), eye movements and emotional function. This area is strongly connected to the cortex and thalamus, along with other brain areas. These areas are implicated in speech and throughout language. Philip Lieberman refers to the disparate parts of the brain that produce language as the *Functional Language System*.[2]

The general nature of the basal ganglia (sometimes referred to as the 'reptilian brain'), their role in speech and their responsibility for habit formation, teaches us several things. First, this region is a fundamental component of language function, even though not specifically evolved for language. It is known that the ganglia are crucial for language because harm to them produces a number of aphasic conditions. If these vestigial portions of the cerebellum and reptilian brain are part of the Functional Language System, however, this indicates that responsibility for language lies with various regions of the brain that contribute in multiple ways at a higher level of organisation in our mental or cortical life than merely language. This tells us that language is at least partially a series of acquired habits and routines, along with others like skiing, bicycle-riding, typing and so on, since habits and routines are the purview of the basal ganglia.

Another reason the basal ganglia are important is that their role in language illustrates the importance of the theory of microgenetics. This theory claims that human thinking engages the entire brain, beginning with the oldest parts of the brain first. Or, as it is put in a recent study:

> The implication of microgenetic theory is that cognitive processes such as language comprehension remain integrally linked to more elementary brain functions, such as motivation and emotion ... linguistic and nonlinguistic functions should be tightly integrated, particularly as they reflect common pathways of processing.[3]

Many researchers underscore why one should not jump to conclusions about the significance of the fact that some kinds of knowledge are found in specific regions of the brain:

> [E]verything humans know and do is served by and represented in the human brain ... Our best friend's phone number and our

spouse's shoe size must be stored in the brain, and presumably they are stored in nonidentical ways, which could ... show up someday on someone's future brain-imaging machine ... The existence of a correlation between psychological and neural facts says nothing in and of itself about innateness, domain specificity, or any other contentious division of the epistemological landscape.

The authors add:

Well-defined regions of the brain may become specialised for a particular function as a result of experience. In other words, learning itself may serve to set up neural systems that are localised *and* domain-specific, but *not* innate.[4]

Therefore, it is very important to exercise care before speculating that any human knowledge is inborn. The brain is built for learning. It is always best to consider learning as the reason for any information found in any part of the brain, at least before claiming it is hardwired knowledge.

It is, of course, possible that there are concepts inborn in humans. But this is a problematic idea. To implant information innately in the brain the human genotype would need to come prespecified as responsible for different concepts, actual propositional knowledge. That is, there would need to be a gene or a gene network for each purportedly innate concept, perhaps something like 'Heights are scary' or 'Don't associate with cheaters' or 'Nouns refer to things' or 'You can't question subjects in subordinate clauses'. On the other hand it is possible that the cytoarchitecture of the brain makes some things easier to learn in different regions due to the types or configuration of cells in those regions or the connections of some regions to other regions. In fact, there is no uncontroversial evidence that brains either do or do not have specialised, hardwired networks or modules independent of learning, aside from their purely physical properties. Notwithstanding this absence of evidence, there are plenty of researchers who insist that concepts are inborn. Related to this, it is believed by some that innate, language-specialised regions of the brain exist. One of the best known of these regions is Broca's area, a hypothesised region on the left side of the brain. (More technically, it is that part of the brain located at the pars opercularis and pars triangularis of the inferior frontal gyrus.)

The purported specialisation of Broca's area was first suggested in the nineteenth century. The claim emerged from the work of the French researcher and physician Pierre Paul Broca, working with a patient he nicknamed 'Tan' because this was the only word that the patient could utter.

To many modern neuroscientists this is no longer as convincing as it was when first proposed by Broca.[5] In fact, for most researchers, Broca's area does not exist as a clearly demarcated part of the brain. As one author further clarifies,

> … anatomical definitions are often quite imprecise with respect to specific language functions that are processed in the cortical areas. Thus, localizing Broca's region in the context of a functional imaging study analyzing linguistic material, or a lesion study of a Broca aphasic may refer to completely different areas with different cytoarchitecture, connectivity and, ultimately, function.[6]

In spite of growing professional scepticism of Broca's studies, many people still think of his subject Tan's affected brain region as a special area for language. Ned Sahin and co-authors have claimed:

> Neighboring probes within Broca's area revealed distinct neuronal activity for lexical (~200 milliseconds), grammatical (~320 milliseconds), and phonological (~450 milliseconds) processing, identically for nouns and verbs, in a region activated in the same patients and task in functional magnetic resonance imaging. This suggests that a linguistic processing sequence predicted on computational grounds is implemented in the brain in fine-grained spatiotemporally patterned activity.[7]

The problem with this type of research methodology, however, is that Broca's area – assuming that one could define its location in any valid way – is more general in its function than language. There are indeed some parts of the brain that are linked to language. There must be, in fact. But they are not generally dedicated exclusively to language. Focusing on language or grammar in a region of the brain such as Broca's area is like claiming that forks exhaust the function of the kitchen.

More precisely, though, today one would say that there are regions of the brain that participate in many cognitive tasks and that they can enter different neural networks for different tasks. For language, the region often spoken of loosely as Broca's area is a part of the aforementioned Functional Language System, linking various multitasking parts of the brain, as needed, for language. Supporting the assertion that Broca's area is not 'the language region' of the brain is the ironic fact that Broca's area can be destroyed without affecting language if the subject is young enough. In other words, Broca's area is not only not exclusively dedicated to language but also regularly engaged during a range of cognitive tasks, such as coordination of motor-related activities.

As an example of other jobs performed by this purported region, consider what happens when someone is shown hand shadows of moving animals – the region near the classic Broca's area is activated. The region is also activated if someone listens to or performs music. But these are clearly not language-specific tasks. Rather, what they indicate is that Broca's 'area' has a function more general than language. It seems instead to be an 'activity-coordination part' of the brain. Language production is all but one activity among many. This is not to say that Broca's area is fully understood or that it is known with certainty that there are no hereditary language-specific regions of the brain. The claim is only that such regions haven't yet been discovered.

Moreover, new evidence being gathered suggests that such areas may never be discovered. Research hints that brains are composed of polyvalent (doing more than one thing) networks, along the lines of the Functional Language System, that can reform or be reused for a variety of diverse functions.[8] Recent findings in research at MIT argue that the 'visual cortex', the region of the brain usually associated with vision in sighted individuals, can be used for non-visual tasks.[9]

Again, this work is extremely important for any attempt to link cognitive functions with specific regions of the brain. Such work is also important for anyone tempted to conflate statements like 'this region of the brain does X, among other things' vs 'this region of the brain is genetically specified to do X and X alone'. These are separate issues entirely. To find something in the brain is not to discover that a cognitive ability is innately specified to be located exclusively in that portion of the brain.

The MIT-led research is by no means alone in showing how impressive brain plasticity is. Genetic transcription factors responsible for the localisation or specialisation of different regions of the brain for different cognition functions do not seem to be the result of genetically determined links between different cognitive functions and cerebral topography. The lungs, larynx, teeth, tongue, nose and so on are all vital for non-signed language, just as the hands are crucial for signed languages, but none of these are either individually or collectively language organs. How bizarre it would be to claim that the hands are language organs.

The same issues arise for any claim of neuroanatomical specialisation for language. Another such area, commonly claimed to be a language-specific area, is a region known as Wernicke's area. This is located in the posterior section of the superior temporal gyrus in the dominant cerebral hemisphere of a given individual. That means that for right-handed people it is found in the left hemisphere. For lefties, language seems more distributed. Though it is still found in the left hemisphere, left-handers have better ability to recover from strokes that affect language due to less narrow localisation for the left-handed than for right-handed folk. At one time, this region of the posterior temporal lobe of the brain was believed to be specialised for understanding written and spoken language.*

Unfortunately for anyone interested in marshalling anatomical support for the innateness of language, Wernicke's area is not exclusively or even mainly dedicated to language. Wernicke's area, just as Broca's area, is part fiction, as there are no agreed-upon definitions of either the location or its extent. That makes it difficult even to say

---

*There is a great deal of discussion in the literature these days about birdsong and its possible connection to language, given the overlap of music in the human brain (see Bolhuis and Everaert (eds), *Birdsong, Speech, and Language*). However, one thing that these discussions all miss is the contribution of culture to human language. As I argue in *Language: The Cultural Tool*, because other animal communication systems, such as birdsong, lack the component of culture as I further define it in *Dark Matter of the Mind*, they cannot have language. And culture can override innate biases, such as they might exist.

that there *is* such an area with any precision. Second, recent research shows that this region is connected to other areas of the brain that are, as was the case with Broca's, far more general in their function than language, such as motor control, including the pre-motor organisation of potential activities – things like getting your fingers ready to play the guitar before you begin to play. Third, as the research above indicates, even if one did find an area specialised for a particular function in one or even in a million subjects, the next subject one meets could in many cases be using that area of their brain differently, depending on their individual developmental history. The lesson to take away from this is that parts of the brain develop in each individual as a home-base for multiple, though related, tasks.

But if the organisation of the human brain is that plastic, how is it that the sapiens brain takes the form and shape that it does? The answer is that brain growth and development is guided not only by genes but also by histones that control 'transcription factors'. A transcription factor is simply a protein that connects ('binds') to specific sequences of DNA. By doing this, these factors are able to determine the rate of transcription. That is how information from the genes is passed from DNA to messenger RNA. These transcription factors are crucial for development. They regulate how genes are manifested or 'expressed'. Transcription factors play a role in the development of all organisms. The larger the size of the genome, the larger the number of transcription factors necessary to regulate the expression of the larger number of genes. Not only that, but organisms with larger genomes tend to have even more transcription factors per gene.

It is also known now that brain specialisation and anatomy can be influenced by culture. This makes it much harder to tease apart evidence for 'pure' biology from biological properties influenced or supplanted by learning or the environment. Psychologists have shown that reading-challenged children who had experienced as little as six months of intensive remedial reading instruction grew new white-matter connections in their brains. And this study is but one of many that have found that culture can change the structure and functionality of the brain. Other studies have shown that the connections between portions of the brain can weaken or strengthen over time, based on the cultural experiences of the individual.

Because culture can change the form of the brain, and since there is no knowledge of any cognitive function that is innate to a specific location in all brains, the difficulty of using arguments from cerebral organisation or anatomy for the idea that language is innate is clear. And it is equally implausible to claim that specific regions of the brain are genetically specialised for specific tasks. The brain uses and reuses its various areas in order to accomplish all of the challenges that modern humans confront. Evolution has prepared humans to think more freely than any other creature by giving them a brain capable of learning culturally rather than one that relies on cognitive instincts. From one vantage point, localisation in the brain is trivial. Everything we know is somewhere in our brain. Therefore, finding that this or that kind of knowledge is located in a specific part of the brain is not evidence for innate knowledge. I was born in Southern California. That doesn't mean I was predestined to be there. Everyone has to be born somewhere.

Occasionally one reads linguistics research claiming that language is housed in the brain and underwritten by specific genes in the same way that vision, growing arms, hearing and our other natural abilities seem to be. But language is not like vision. Vision is a biological system. Language is perhaps a bit like the use of vision in perception, in that it requires culture for interpretation and use (as in art and literature). But we now know that language is a nexus of social, computational, psychological and cultural constraints and requirements. With experience, as we age and learn, parts of the brain come to specialise for and house components of language. But this is true for everything we know. I know how to boil water. That is somewhere in my brain. But neither boiling water nor language is innate simply because it is found in a specific part of the brain – not even if such things are found in roughly the same part of all brains across all individuals.

It is also often claimed by Rutgers philosopher Jerry Fodor and others that language is an encapsulated mental module (meaning that it operates independently from the rest of the brain). But Evelina Fedorenko, an MIT neuroscientist, has shown that when we use language, we always draw on both specific and general knowledge.[10] First, an individual might access a particular word meaning stored in his or her brain, but they will subsequently also access the general cultural knowledge they possess in order to interpret that word in their present circumstances. Therefore, language is not encapsulated, nor is it an autonomous ability. And it does not have a genetically dedicated home in the brain. But innate location and autonomy are often appealed to in order to claim that language itself is innate, an encapsulated module.

Every neuroscientist worries about whether their own brain is up to the task of understanding the human brain more generally. Some, like Ramón y Cajal in the quote at the beginning of this chapter, think not. Others believe that such pessimism is unwarranted and that one must press on, making gradual progress, as with any difficult object of study. Optimists believe that full understanding of the brain can be achieved homeopathically – a little at a time, learning a tiny bit more about the brain experiment by experiment, as everything in science is learned. The banal fact that the brain is necessary for language doesn't shed much light on how the brain functions in language.

For some of these reasons, interest in the brain has grown tremendously over the past fifty years. In 1970 the Society for Neuroscience was formed, with 500 founding members. As of this writing, the society has more than 35,000 members worldwide and holds conferences with an average audience of 14,000 for presentations and a total of 30,000 attendees. The philosophy of neuroscience is another recently popular field that, according to most of its practitioners, originates in the 1986 book *Neurophilosophy* by Patricia Churchland. There is tremendous variation in beliefs, theories and research interests in both of these fields of study, but one of the emerging points of consensus among many (though by no means all) is that brains are simply one organ of the entire individual's physiology and that cognition, activity, ascriptions of abilities and so on are properties of the entire individual, as well as the culture in which that individual is found.

Part of the reasons for neuroscepticism, the belief that humanity will never understand the human brain, is expressed in this passage:

Gram for gram, the brain is far and away the most complex object we know of in the universe, and we simply haven't figured out its basic plan yet – despite its supreme importance and a great deal of effort ... No Mendeleyev, Einstein, or Darwin has succeeded in grasping and articulating the general principles of its architecture,

nor has anyone presented a coherent theory of its functional organisation ... there is not even a list of basic parts [of the brain] that
neuroscientists agree on.[11]

What evolved in the skull from the earliest primates to *Homo
sapiens* is an organ of the body. The brain is not, nor does it contain,
an ethereal entity such as a mind or a soul. The brain is nothing less
nor more than the main organ of the nervous system, as the heart is
the main organ of the circulatory system, or the lungs are the main
organ of oxygenation, or the nose is the main organ of smell, or the
eyes the main organs of vision. The brain cannot live or develop on its
own. It is, like any corporeal organ, shaped and constrained by all the
other physiological systems of the body, in cooperation with cultural
experiences, individual apperceptions, the food we eat, the exercise we
take and, generally, just the way we have lived. The brain is found in
the head, which itself has evolved to fit and protect it, changing as the
brain changes and essential to its proper functioning.[12]

Much of what is known about the human brain has been learned
from experimentation on the brains of live animals. For human brains,
less-cruel methods are used, such as functional neuroimaging of different types and electroencephalogram (EEG) recordings. These more
humane methods of brain study have provided a multitude of insights
into how the brain supports human language.

The brain's three pounds consist of neurons, glial cells and blood
vessels. Each plays a vital role in proper brain functioning, intelligence
and other cognitive abilities of the species. There are some 100 billion
neurons in the average brain. There are also non-neuronal cells of a
roughly equal amount. Nearly 20 per cent of all neurons in the brain
reside in the cerebral cortex, including white matter found beneath the
cortex or 'subcortical white matter'.

The biggest part of the brain is the cerebrum (from the Latin word
for 'brain'), which sits underneath to the rear of the cortex (literally,
the 'bark' of the brain). The cerebrum is broken into two hemispheres.
When someone talks about being 'left-brained' or 'right-brained', they
are referring to the two hemispheres of the cerebrum.

As we can see from the ventral (underside) view of the brain shown
in Figure 18, what in turn sits under the cerebrum is the brain stem.

Behind the brain stem is the cerebellum. Nearly 69 billion of our brain neurons, 80 per cent of all our neurons, are located in the cerebellum or 'little brain', which sits just below the cortex.

The cerebral cortex is convoluted (lots of ridges and valleys), a common feature of larger brains, regardless of species. The brain is soft and, were it not for the skull housing it, would be very easy to damage. Many components of the human brain are found in all vertebrate brains. These include the medulla oblongata, pons, optic tectum, thalamus, hypothalamus, basal ganglia and olfactory bulb.

The major human brain divisions of forebrain, midbrain and hindbrain are likewise common throughout the animal kingdom. Mammalian brains are more advanced than the average vertebrate brain, however. All mammals have a six-layered cerebral cortex. As primates, humans have larger cortices than the brains of non-primates and, like other primates, their brain shape has been mildly affected by the fact that they hold their heads upright.

But humans are not merely vertebrates, mammals and primates – they are hominins, possessors of the largest brain relative to their body size in perhaps the entire animal kingdom. But in particular they possess the densest packing of neurons of any creature. Human brains are formed through gene-environment interactions. There are barriers (some of blood and others of cerebrospinal fluid) around the brain and specialised cells (glial cells and mast cells) that offer an independent neuroimmune system, distinct from the immune system for the rest of the body. Humans share much of our brain structure with other hominins, such as *Australopithecus* and *Homo erectus*, though sapiens brains are larger, more specialised and more complex than those of other hominins.

If one were to flatten out the folds of an entire anatomically modern human's cortex on the floor, it would have a surface area of approximately 2.6 square feet, or $0.24m^2$. The folds of the human cerebral cortices have raised and lowered portions, ridges and grooves, each called a gyrus (a raised portion, from the Greek for 'ring') or a sulcus (an indented portion, from the Latin for 'furrow' or 'wrinkle').

Each of the cerebral hemispheres is divided into four lobes (Greek for 'pod'): frontal (Latin 'front'), parietal (Latin 'walls of house'), occipital (Latin 'back of head') and temporal (Latin 'lasting for a time').

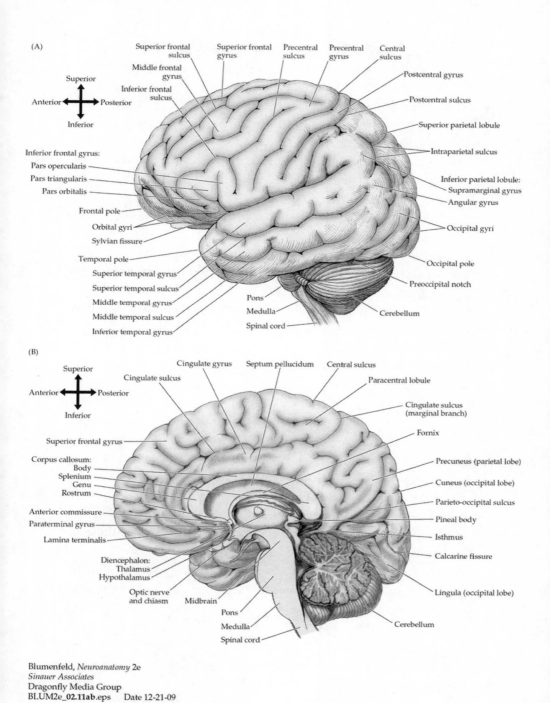

Blumenfeld, *Neuroanatomy* 2e
Sinauer Associates
Dragonfly Media Group
BLUM2e_**02.11ab**.eps   Date 12-21-09

Figure 17: **Midsagittal view of the brain**

Figure 18: **Ventral view of the brain**

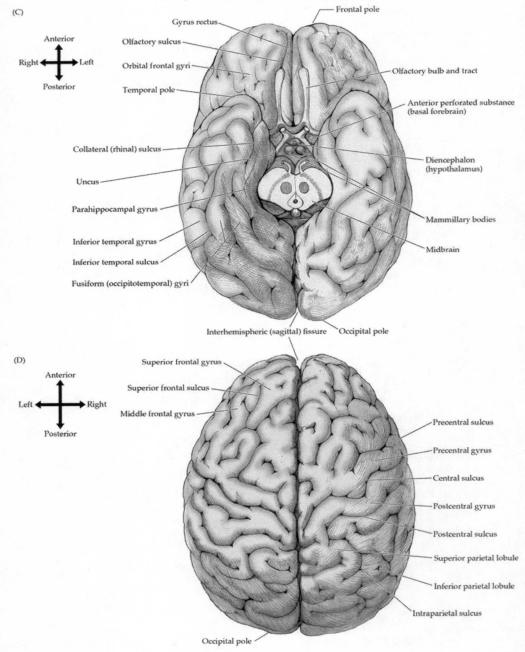

(C)

Anterior
Right ←→ Left
Posterior

Frontal pole
Gyrus rectus
Olfactory sulcus
Orbital frontal gyri
Temporal pole
Olfactory bulb and tract
Anterior perforated substance (basal forebrain)
Collateral (rhinal) sulcus
Diencephalon (hypothalamus)
Uncus
Parahippocampal gyrus
Mammillary bodies
Inferior temporal gyrus
Midbrain
Inferior temporal sulcus
Fusiform (occipitotemporal) gyri
Interhemispheric (sagittal) fissure   Occipital pole

(D)

Anterior
Left ←→ Right
Posterior

Superior frontal gyrus
Superior frontal sulcus
Middle frontal gyrus
Precentral sulcus
Precentral gyrus
Central sulcus
Postcentral gyrus
Postcentral sulcus
Superior parietal lobule
Inferior parietal lobule
Intraparietal sulcus
Occipital pole

Figure 19: **Dorsal view of the brain**

These are named in the main for the corresponding portion of the skull immediately above them. One sees references to these lobes frequently in the neuroscientific literature, but their primary purpose is to refer to a general region. They are too poorly understood and demarcated to attribute to them any significant function in the overall behaviour of the brain. In fact, different brain functions are commonly distributed across the supposed borders of the different lobes. An added complication is that every human brain is unique – no two brains have exactly the same pattern of gyri and sulci. Yet human brains all function similarly, in spite of gyri and sulci patterns. This indicates either that the functions of brain folds are not fully understood, or that the brain folds make little difference to brain functioning, or that there are brain characteristics that distinguish individuals in ways not yet identified. All of these options are in all probability true in some sense.

However primitive our current understanding of it, it is clear that the brain's anatomical architecture is very important to language and other cognitive functions of our species. But there is another kind of architecture beyond the gross anatomy which is equally important, if not more so, to brain functions: its cytoarchitecture.

In the midsagittal view of the brain (Figure 17), the brain stem, the cerebellum and the cerebrum are all visible. What is worth noting in particular in this diagram, however, is the 'lie of the land' and the fact that specific regions of the brain are where we most commonly find the control centres of certain physical abilities. The fact that these abilities are found in most subjects does not necessarily mean that they cannot be repurposed for other uses. The cognitive scientist Elizabeth Bates and her colleagues in fact developed a model of brain specialisation that relied on the order and manner in which things were first learned and the physical nature of the relevant cytoarchitecture, avoiding in many cases the need to call upon unidentified genes. This is not to say that genes do not play a role in the architecture of the brain. Of course they do. But they should not be invoked without understanding their role. Two other views of the brain that illustrate its overall complexity are the ventral (lower) and the dorsal (upper) views (Figures 18 and 19).

More detailed representation of the cognitive functions of brain areas come from 'cytoarchitecture'. This is the cover term for the

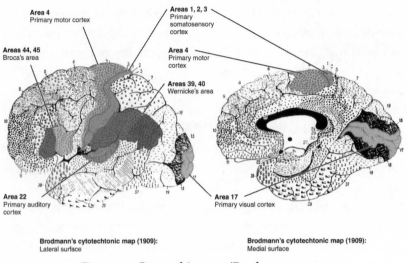

Brodmann's cytotechtonic map (1909):   Brodmann's cytotechtonic map (1909):
Lateral surface   Medial surface

Figure 20: **Cytoarchitecture/Brodmann areas**

differences in the construction of individual cells found in particular
brain regions. The division of brain regions based on this cytoarchitecture is given in the Brodmann classification of areas of the cortex
distinguished by cell forms (Figure 20). This organisation is manifested
in various distinct ways – connections between cells, shapes of cells or
cell parts, thickness of the cortex in a particular cellular region, according to which region of the cortex we are viewing and what function that
region has. It is named after Korbinian Brodmann, the first to propose
this type of division of the brain. Most of the divisions he proposed are
linked to parts of the body or higher cognitive functions.

Based on the cytoarchitectural division of the brain into Brodmann areas, recent proposals searched for specific linguistic regions
of these areas that might explain language's uniqueness to humans.
Some of these proposals assume a view of the neurological basis for
language of *sapiens* that is somewhat problematic. With all proposals
on the anatomy or evolution of the brain, one should be suspicious of
attempts to interpret neurology as being either predicted by or supportive of highly specific theories of language. In particular, Brodmann's
areas offer no direct support to the 'X-Man' view which takes language
to be a special, saltational change resulting from a mutation.[13] Considering this proposal and rejecting it paves the way for a discussion of
more general properties of the brain necessary for language.

The proposal is partially built off the fact that there have been multiple evolutionary innovations in the human brain. Thus some researchers have claimed that there are differences in the sapiens brain resulting from a mutation for grammar. The mutation proponents also claim that no other species past or present, such as *Homo erectus*, could have had language. This is because such researchers claim that there is no evidence for symbolic representations among *Homo erectus* or, probably, even *Homo neanderthalensis*. Therefore, only *sapiens* could have had language. This is a doubly erroneous idea. The two problems with it are clear enough. First, *erectus did* have symbolic representations – tools, status symbols and etchings. Second, the method is wrong. It entails a claim based on the absence of evidence, namely that symbolic art is not found in non-*sapiens Homo* species. But absence of evidence is a weak argument. And in this case there is evidence. Possibly third, this idea that *erectus* lacked language ignores their accomplishments. Remember that *erectus* sailed. This activity alone demonstrates that *erectus* was able to think ahead, to imagine and to communicate. Even if *erectus*'s language lacked hierarchical syntax, this would differ only in degree and not in kind from modern languages. In other words, the language of *erectus*, what might be called a 'G_1' (early grammar) language, could evolve seamlessly into a contemporary 'G_3' (later grammar) language – one with recursion and hierarchy. And after the G_1 language was invented by *erectus*, and quite possibly adopted subsequently by *neanderthalensis* and *sapiens*, G_3 arrived after little more than a few years of tinkering (could be thousands of years or dozens of years or only a few years). Some researchers have correctly demonstrated that the evidence for language goes back at least to *Homo neanderthalensis*, some 500,000 years ago. But there is no reason that language could not reach back even further. Like *sapiens*, *neanderthalensis* were beaten to the invention of language by their ancestors, *Homo erectus*.

Some problems attendant to the proposal that the brain has specific encapsulated modules dedicated to language can be seen by briefly reviewing research reported by neuroscientist Angela Friederici of the Max Planck Institute for Human Cognitive and Brain Sciences in Leipzig. Friederici has written extensively on her research on hypothesised brain mechanisms underlying language. She adopts assumptions on the nature of language as a specific type of complex grammar. The claim of hers worth scrutinising here is the proposal that Brodmann Area 44 (BA 44) is a functionally specific (that is dedicated to a specific task) range of tissues (BA 44 is visible in Figure 20). Friederici claims that this area is connected to the temporal cortex and that this is the neural difference that triggers or enables the capacity of *Homo sapiens* for syntax. Before examining this bold claim, a simple question is in order. What did the researcher believe about their subject before beginning their research? This question can be helpful for detecting the early signs of 'confirmation bias' (obtaining experimental results that confirm one's pre-existing beliefs).

This particular research begins by assuming the narrow view that language is a grammar with recursion. No communication system lacking this property can be a language, according to this perspective. This assumption is made in many experiments on the cerebral support for language. In many cases the researchers assume a grammatical rule called 'Merge' as the basis for all human language. Merge is simply an operation that combines two objects to form a larger object. Assume that one wishes to utter or interpret the phrase, 'the big boy'. This phrase is not formed by simply placing the words 'the', 'big' and 'boy' together. In the Merge theory of grammar one takes 'big' then combines that with the word 'boy'. Then 'big boy' is combined with 'the'. As it is designed, the rule of Merge cannot work with more than two words or phrases at a time. This means that it, and therefore all of language is, a 'binary' procedure. This is a conception of language as structure, rather than as meaning and interaction.

In making the assumption that human language = Merge, researchers arguably overestimate the importance of syntax, which plays only a minor role in human language as a means to organise information flow. As such, it acts as a filter, to help guide the hearer to the intended interpretations of an utterance.* Grammar is symbols used together. Syntax is the arrangement of those symbols as they are used together. Compared to the invention of symbols and the cultural basis of their meaning, however, along with the knowledge of how to use symbols appropriately in telling stories, conversing and using language in its

---

*The use of syntax as a filter was first proposed by Chomsky himself in his 1965 book, *Aspects of the Theory of Syntax* (Cambridge, MA: MIT Press, 1965).

many forms, syntax is a helpful tool, but arguably non-essential. The problem is that the operation Merge is not found in all human languages. In several modern languages it has emerged that there is no evidence supporting a binary syntactic operation.* Merge is therefore not a prerequisite for human language. Moreover, even if it were, Merge is not specific to language. It is an example of the well-known process of associative learning, of the type that made Pavlov's dog famous. Pavlov's pooch learned to associate a ringing bell with food. The bell was rung just as the food was served. Eventually the dog 'merged' these two concepts, bell and food, and would salivate at the sound of a bell.

Yet if Merge is simply a form of associative learning, then it cannot be linguistically unique. Associative learning is commonly found both outside of syntax and language and across most species. Arguably, all animals possess a capacity for associative learning. So this is the first problem with Friederici's claims. She has the wrong concept of syntax.

But the conceptual neurolinguistic problems don't stop there. It isn't just the erroneous concept of syntax adopted that damages attempts to show language specificity in the brain. The main problem is not merely the idea that there is innate language that is a simple syntactic operation, but the idea that one could locate an innately hardwired place for language in the brain. That idea is like asking the brain, 'Where's the syntax?' Well, of course, syntax is in some place or places in the brain, yet just because syntax is found in the brain does not mean that it is genetically prespecified to be there. Moreover, it is possible that some areas of the cytoarchitecture of the brain are indeed propitious for storing or operating syntax. But that is not the same as saying that BA 44 and its temporal lobe connection are the evolutionary developments that pulled language out of a primate brain. BA 44, like any other region

---

*For example, Merge makes the wrong predictions about the Wari' language of the Amazon (see Daniel Everett and Barbara Kern, *Wari'* (London/New York: Routledge, 1997)). Linguists Ray Jackendoff and Eva Wittenburg have claimed that one looks in vain for Merge in the Riau language of Indonesia ('What You Can Say Without Syntax: A Hierarchy of Grammatical Complexity' – https://ase.tufts. edu/cogstud/jackendoff/papers/simplersyntaxwritten.pdf). And Jackendoff and renowned syntactician Peter Culicover co-authored a book entitled *Simpler Syntax* (Oxford University Press, 2005) in which the authors claim that not all languages use the same syntactic operations.

of the brain, has a broad range of functions and is not exclusively associated with syntax. BA 44 has at least six separate functions, including
sound or phonological processing, syntactic processing, understanding meaning or semantic processing, and music perception. It is also
employed in responding to 'go' vs 'no go' decision-making. BA 44 is
also used in hand movements. So BA 44 is not 'for syntax'. It might be
necessary for syntax, but that's like saying that the hand is necessary
for pencils. The different roles of BA 44 illustrate that we need a more
general, higher-level understanding of what it does. We cannot call it
a language organ.

Work like Friederici's errs not only because it assumes that syntax is
more important for language than it in fact is but also because the relevant portions of the brain are not as specialised as this work assumes.
Researchers of this persuasion also seem to suggest that language is
simpler and younger than the evidence we have reviewed indicates.
Not only is language much more complex than research like Friederici's assumes but, if I am correct, it is also significantly older. Nearly
2 million years older. It predates perhaps even the evolution of relevant
portions of the brain that she claims to be crucial for language.

This does raise an interesting question for the relationship between
the brain and language, though. If language is indeed as old as it seems
to be, then did *Homo erectus* or *Homo neanderthalensis* also have BA 44?
That is, can language exist without the modern brain area known as BA
44? No one knows. But then if there is no idea whether this area is found
in other species of *Homo*, no one has any idea which parts of the brain
supported syntax in *erectus* or *neanderthalensis*. This means that it is
not known whether BA 44 is necessary for syntax. One suspects that,
because the human brain is flexible, different parts of it will be exploited
at different times in the evolutionary history of our genus subsequent to
erectus. The portion of the brain used by modern humans may be simply
the neurobiology that contemporary *Homo* brains avail themselves of.
Earlier species may well have used different brain structures for syntax.

Another issue with Friederici's analysis of BA 44 has to do with the
fact that it was inspired by very famous experiments with cotton-top
tamarin monkeys that were conducted by Tecumseh Fitch and Marc
Hauser in the earlier part of the 2000s. The experiments are problematic for at least two reasons.

The first reason is that they may be unreliable. I watched an unsuccessful attempt to run a version of these experiments by Fitch among the Pirahãs in 2007 when he was my guest in the Amazon. But if these experiments failed with human subjects, it is possible that they also did not run as well as believed among tamarins. Perhaps they are not sensitive to syntax in the way claimed for them.

The most serious problem with these experiments, however, is that the science behind them is flawed. In the *Psychonomic Bulletin and Review* this response was published by Pierre Perruchet and Arnaud Rey:

> In a recent Science paper, Fitch and Hauser ... claimed to have demonstrated that Cotton-top Tamarins fail to learn an artificial language produced by a Phrase Structure Grammar ... generating center-embedded sentences, while adult humans easily learn such a language. We report an experiment replicating the results of F&H in humans, but also showing that participants learned the language without exploiting in any way the center-embedded structure. When the procedure was modified to make the processing of this structure mandatory, participants no longer showed evidence of learning. We propose a simple interpretation for the difference in performance observed in F&H's task between humans and Tamarins, and argue that, beyond the specific drawbacks inherent to F&H's study, researching the source of the inability of nonhuman primates to master language within a framework built around the Chomsky's [sic] hierarchy of grammars is a conceptual dead-end.[14]

Adding to the scepticism about Fitch and Hauser's results, Professor Mark Liberman offered his own response, on the widely read blog *Language Log*, concluding that Fitch and Hauser's findings were in all likelihood about memory, not grammar per se.[15] But if that is correct, then these studies have no bearing on Friederici's claims, offering no support for either her methodology or her conclusions.

What this scepticism shows is that the popular idea that human brains are hardwired for language is not confirmed by science, even though it is often claimed. Nevertheless, there has always been a temptation since brain studies began in earnest, to associate the structure of the brain – its lobes, layers, sections and other gross anatomical features – with different kinds of intelligence or distinct tasks. The outmoded 'science' of phrenology or localisation was a consequence of trying to associate physical features of our skull with the cognitive, emotional and moral properties of the brain encased within it. This is another error of overusing the Galilean metaphor of the universe or the brain as a clock.

The human brain must be able to follow conversations, use words appropriately, remember and execute pronunciations, decode pronunciations it hears from others, keep track of the stories in the conversations, remember who is being talked about and follow topics through long discussions. And this is only a small list of the ways in which language requires memory. No memory, no language. No memory, no culture. But language requires a special set of memories, not just any memory. The different varieties of memory that underlie language are sensory memory, short-term (or 'working') memory and long-term memory.

Sensory memory holds information from the five senses in the brain for very short periods of time. It is able to capture visual, aural, or tactile information in less than a second. In this way sensory memory enables one to look at a painting, say, or hear a song, or feel someone's touch and remember what the experience looked like, sounded like, or felt like. This type of memory is vital to learning from new experiences. And it is particularly important for language learning – remembering how new words sound long enough to repeat them and build them into long-term memory. Sensory information is like a reflex and just seems to happen, but is not sufficient alone to support language. It degrades very quickly. If one examines a sequence of numbers, they might be remembered (though sensory memory is limited to a maximum of about twelve items), but in all probability not long enough to walk across the room and write them all down. There are three varieties of sensory memory: *echoic*, for sounds; *iconic*, for vision; and *haptic*, for tactile sensations.

Another kind of memory, short-term or 'working' memory, is also crucial in the use of language. At an MIT conference on 11 September 1956, remembered for a series of brilliant lectures that some refer to now as the 'cognitive revolution', psychologist George A. Miller, then of Bell Laboratories, later of Princeton, presented a paper entitled 'The

Magical Number 7 +/– 2'. Miller's research concluded that, without practice, people can remember up to nine, usually more like five, items at a time for roughly a minute. Some have come to disagree with Miller and believe working memory is actually lower than this, around four items at a time. Miller discovered, however, that if items are 'chunked', then people are able to remember larger numbers of items. This turned out to be a great result for science, but also for Bell Labs. Bell discovered that a person might have difficulty remembering the number 5831740263 but that they could easily remember it if it were 'chunked', as in (583) (174) (0263). It further turns out that working memory is biased towards sound-based memories, which means not only that it is important for remembering and decoding utterances, but also that it seems to have partially evolved for that very purpose, language once again helping to shape human evolution.

The next form of memory is long term. Most people remember a great deal of their childhoods. The memories may be partially inaccurate in some ways, altered over time by different conversations about them, but something of those early experiences is there for the entirety of people's lives. Long-term memory can recall vast amounts of data for a nearly unlimited period of time within the confines of human lifespans.

Long-term memory is divided into *declarative* memory and *procedural* memory. Procedural memory is implicit memory of processes involving motor skills. When trying to remember a password, your declarative memory might fail you, in the sense that you cannot recall the names of all the symbols you have chosen for your password. But your procedural memory can come to the rescue if you will simply set yourself down at the keyboard and type the password. In a sense, your fingers 'remember' a code that your conscious memory has forgotten. Or someone might be trying to teach another person how to play a riff on the guitar and forget the notes. They can still teach the riff, however, by slowly playing it for the learner. But be careful not to play too slowly. Procedural memory seems to prefer to keep things at tempo.

Procedural memory is vital for pronunciation or for gestures in sign languages, providing much quicker use and access to words and signs than declarative memory, just as one's fingers might better remember their computer password. Without procedural memory, language and much of culture would be impossible. Obviously, any animal that can do routines quickly also has a form of procedural memory. It is not unique to humans.

Declarative memory is subdivided into semantic memory and episodic memory. Semantic memory is associated with facts independent of any context, such as 'a bachelor is an unmarried male'. It is crucial for linguistic meanings. And it is also vital for other context-free, long-term memories, such as '9/11 was a horrible day'.

Episodic memories, on the other hand, are long-term memories associated with a specific context and, therefore, tend to be more personal. You might use this memory to recall, 'That's where we had our first water pistol fight.' Or, 'That's the Mexicali bar where I had my first tequila.'

Working memory takes place as exchanges between neurons in the frontal cortex. But long-term memories are widely distributed in the brain and seem to be processed initially by the hippocampus, which 'consolidates' memory for long-term storage elsewhere. Thus we must have brains that can support the three basic types of memory humans use so frequently and depend on to survive and to speak.

Several cultural traditions have attributed various functions to specific bodily organs and their parts, such as emotions being centred in the heart, thoughts being processed in certain parts of the brain and language being rooted in others. But culture can only get things right as its scientific methods, output and understanding evolve over time. It is now almost universally accepted that the heart has nothing to do with emotions: it is a blood pump, nothing more. And it is likewise, though perhaps less widely, now known that, although the brain is fundamental to the processing of all of our cognitive functions, it is not the sole locus of those functions. The resources of our entire body are marshalled in thinking, just as they are in communicating. (If you doubt this, imagine how illness or a hangover can affect your ability to think clearly.) And the brain's thoughts largely come from the totality of our personal experiences as they store and embellish them. These are also known as apperceptions – the experiences that make people who they are. But let's call the question. How is it that the brain enables people to speak? And what prevents other animals from having language?

The French philosopher René Descartes was the morning star of the Renaissance, a touchstone in the culture and history of the Western world. He was a pioneer who brought innovative, original thinking back into fashion after more than five centuries of talking about people instead of ideas. Before Descartes and a few others, there yawned back into the shadows the nearly 1,000-year-long, oppressive Dark Ages, during which 'reasoning' was an act of power. Because of this, ad hominem arguments for one's views – arguments based on some person's reputation rather than the ideas at the heart of their arguments – were the standard.

Descartes's work on the mind revolves around his popular thesis of dualism – the proposal that the mind is our soul and is non-material, be it spiritual, Platonic, or merely mental, while the body is material. According to Descartes, these are two irreconcilable substances. He suggested that the soul and the body were connected, however, through the pineal gland. For whatever reason (perhaps it is due to religious traditions that oppose the soul and body), dualism has been an influential thesis for more than 400 years. Descartes's work was the basis for much of Noam Chomsky's theory of the mind and language, and their relationship, as Chomsky explains in his book *Cartesian Linguistics*.[16] Chomsky seems to support Descartes's claim that, while the body is a machine, the mind is not obviously physical.

Dualism places evolutionary accounts of human thinking at a disadvantage, for the simple reason that something non-physical could not have evolved. Therefore, if we accept this dualism, we reject the idea that the mind evolved by natural selection. In public talks and written works one often sees cited the erstwhile theist Alfred Wallace, co-discoverer of natural selection, as support for the dualistic view of the mind as somehow a different substance from the body. Wallace did indeed believe that the mind, as a Cartesian non-material entity, could not have evolved. It would seem, however, that the best way forward is not the way of dualism but the simpler idea that one should first try to explain things in natural, physical terms before proposing new entities or substances. This would especially apply if one believed that there was no physical basis for thought. To argue for such an idea, one would need to begin with the mundane evidence of the fossils, of DNA, of theories of language and culture, as well as comparative primatology, before proposing non-physical explanations for language, human intelligence, or human minds more generally.

Theories of the human brain and mind have been around for millennia. Scholars often are careful to distinguish these two terms, brain vs mind, but this separation is ultimately inspired by both dualism and religion. In this conception, the word 'mind' refers to the brain activities and properties people are currently unable to explain in physiological terms. But this position assumes that science could in principle eventually provide neurophysiological explanations for the properties people today refer to as the mental.[17]

Earlier it was mentioned that palaeoneurologists practise a reverse form of phrenology, when they study endocasts of fossil skulls (the inside of the skull rather than the outside). Some endocasts are naturally formed as the skulls of dead creatures fill with material that fossilises and is left with physical impressions from the inside of the skull. Other endocasts are made by the palaeoneurologist. Researcher Ralph Holloway has summarised how this is done.[18] The procedure is to first fill the cranium with layers of latex. Next, when the latex is roughly 1–2 mm thick, place the skull in an oven for three to four hours to cure it, after which you force the latex out of the brain cavity. Holloway says that this can be done by collapsing the latex, using baby powder to prevent it sticking to the fossil, and then carefully extruding it from the fossil through the foramen magnum. After the latex is removed, it will revert immediately to the form it had in the skull.

There is nothing wrong with the reverse phrenology of endocast reading, so long as the palaeoneurologist remembers that the brain regions identified on the endocast are only suggestions for what might have been, not unambiguous evidence for, certainly not proofs for or against, different cognitive or linguistic abilities. Endocasts are less informative than one might like because they say so little about the fossil's cytoarchitecture, about white- vs grey-matter distribution and about relative ratios of cell types, such as glial cells to neurons and other aspects of brain anatomy, such as neuronal density, that are never revealed on the underside or outside of skulls.

Another problem is that the brain is in some ways more like a blob and less like a heart, with its separate chambers. In the brain different tasks are performed by spontaneous or pre-existing connections that draw first on the parts of the blob that are most conducive to the task and subsequently on more and more firepower until the task is done. Many of these initially spontaneous connections become more routine after frequent exposure to similar tasks, indicating that learning has occurred. There is a system to the activations, but that system is less anatomical than electrochemical. It is more fluid and dynamic than static and hardwired.

In brain organisation, chemicals rule. Hormones generated by our emotions, thinking processes, diet and the overall state of our entire organism control our brain. This is why many neuroscientists have embraced the theory that the brain is 'embodied' – built into an anatomical, chemical, electrical and physically constrained system, namely our bodies. To such researchers, it is not so much that the brain thinks as that the entire individual does. Thus the brain is a physical organ, a constituent of the body, as all other organs are. This embodiment, along with the role of culture in our thinking, means that the brain is an organ physically integrated into the world through a body, and not a computer.

The picture of the brain that is emerging here, then, is of a cognitively non-modular organ with no congenitally specialised tissue for language (or cooking or guitar-playing). This is in direct contrast to the innately specialised areas that exist for physical abilities, but not for cultural or conceptual ones. If it is correct that language is a cultural artefact, the absence of a specialised brain area for it is predicted. If this idea is wrong, though, then language is more like vision and there should be evidence that language is innately linked to a particular location of the brain, specialised for language.

Further evidence for this general-purpose image of the *Homo* brain comes from disorders of language and speech. It turns out, very surprisingly, given claims frequently found in the literature, that there are no heritable language-specific disorders, supporting the non-compartmentalised theory of the brain.

7

When the Brain Goes Wrong

… language disorders do not occur in isolation; aphasic disturbances rarely occur in the absence of memory impairment or attention/executive problems.

Yves Turgeon and Joël Macoir[1]

ONE WAY TO TEST the prediction that language is innate and hard-wired in certain portions of the brain is to examine the nature of language disorders. If language is an encapsulated, innate module of the brain, then it should be possible to observe language problems that are linked exclusively to particular language-only parts of the brain. If, on the other hand, language is a culturally acquired invention, there should be no specifically linguistic disorders. While the latter idea, that there are no heritable, language-only disorders, seems correct, there is a list of respected researchers that claim otherwise. To help determine what the truth really is, the evidence from so-called language-specific disorders needs to be considered here.

The place to begin is the disorder with the challenging label, specific language impairment (SLI). SLI supposedly affects linguistic functions exclusively. No other part of the brain or other aspects of our cognitive functioning is purported to be affected in this syndrome, according to some researchers. Some assert that SLI shows that the brain is genetically hardwired for specific linguistic knowledge, because this eponymous deficit affects only linguistic knowledge.

In fact, this label is misleading because it suggests that something has been found that in fact has never been discovered, namely a disorder that affects only our linguistic abilities. To the contrary this disorder always seems to affect non-linguistic aspects of our cognition as well. Therefore, whatever the nature of such impairments are, they are not language-specific.

Even if SLI existed as described, this would not indicate that there is a portion of the brain hardwired by the genes for language. That is because acquired skills and knowledge can also be affected in addition to language. Factors such as amnesia, blunt force trauma, alcohol and bullets are some of the causes of language disorders. Therefore, the existence of a disorder implies on the surface nothing about whether the ability affected by the disorder is innate or learned. On the other hand, the model that takes language as an innate capacity of humans does predict highly specific language deficits. The view that language is instead an invention, a cultural artefact, predicts that language deficits are no more likely than bread-baking deficits.

And it turns out that the thesis that the brain is a general purpose device predicts exactly what is found. Deficits that affect language are multifaceted. The language effects themselves are never more than a portion of the overall syndrome. Therefore, a more detailed look at a few so-called language deficits is called for. Michael T. Ullman and Elizabeth I. Pierpont, neuroscientists at Georgetown University, define SLI along the following lines:

> Specific Language Impairment (SLI) is generally defined as a developmental disorder of language in the absence of frank neurological damage, severe environmental deprivation, or mental retardation … Other terms have also been used … such as developmental dysphasia, language impairment, language learning disability, developmental language disorder, delayed speech and deviant language.

There are researchers who consider SLI a dysfunction or deficit of the 'language module' of the brain, the mental capacity exclusively responsible for language. Several explanations of SLI have therefore been phrased in terms of syntax (sentence structure), or phonology (sound structure), or morphology (word structure). These are well-defined submodules or subcomponents of language in some theories. One prediction of SLI might be that a child with this disorder is unable to construct the right types of syntactic tree diagrams mentally, however such structures are implemented in the brain. A linguist might predict that someone could be born without the ability to perform the syntactic operation Merge.

An alternative analysis of this ailment is proposed by Ullman and Pierpont, namely that what is referred to as SLI should instead be understood in terms of a 'procedural deficit'. This counterproposal claims that 'a significant proportion of individuals with SLI suffer from abnormalities of this brain network, leading to impairments of the linguistic and non-linguistic functions that depend on it'.[2] Moreover, they conclude that Broca's area, the region of the brain associated with not only grammar but most procedural motor skills and activities, is implicated, along with procedural memory (our ability to remember how to do things in sequences), meaning that although one sees effects in language, the root of SLI is not linguistic per se, but something of a higher order, the memory and procedural learning that underlie many human activities, of which language is but one. In the words of Ullman and Pierpont:

> The different structures of the procedural system provide distinct and complementary computational and functional contributions. For example, the basal ganglia are particularly important for learning new procedures, but may be less so for the normal processing of already-learned procedures ... So, abnormalities of different structures in the system should lead to different types of impairments of procedural memory.[3]

Although there are many studies such as the one mentioned here against the existence of SLI, it would be premature to consider the matter settled. But, as of this writing, it can be concluded that SLI offers no support for the idea that grammar or language is innate as opposed to the idea that each is a cultural tool that is malleable, changes over time and is learned afresh by every child as they engage in natural conversations and interactions with other members of their community.

Nevertheless, in a 2014 article in the journal *Cell*, Steven Pinker and Heather van der Lely take a very different view of SLI. They claim that SLI is 'highly heritable' and very specific to language. Heritability of SLI is an interesting observation. It would mean that the deficit comes from some hardwired, genetically preordained ability that can become a disability of malfunctioning genes.

But other work has argued that the heritability of SLI may be little more than a gleam in the eye of the beholder. How you define a disease can determine what you find. One study of this kind concludes that: 'Heritability estimates for specific language impairment (SLI) have been inconsistent ... a recent report from the Twins Early Development Study found negligible genetic influence.'[4]

Moreover, what is inherited is wider than merely a language disorder. Other aspects of cognition are affected. This means that the genetic problem is not so much linguistic as a more general problem of brain processing. The evidence that SLI is a more general malady is just what would be predicted if there were no brain module for language, as the evolutionary record seems to suggest.

As others put it:

> The choice of a particular assessment method, the selection of evaluation tools as well as the interpretation of results are highly dependent not only on the clinician's own conception of language but also on the reference to an assessment model ... language disorders do not occur in isolation; aphasic disturbances rarely occur in the absence of memory impairment or attention/executive problems ... Language production and comprehension are complex cognitive skills which should not be considered in isolation in assessment procedures.[5]

Another problem that affects human language is a complete or partial inability to speak, referred to as *aphasia*, from the Greek *a* 'without' and *phasis* 'speech'. Aphasia is caused by brain damage and regularly affects about one million people in the USA alone. Its primary cause seems to be a problem with blood flow to the brain, which is often the by-product of a stroke.*

---

*One of the best descriptions of the disaster of aphasia is discussed in *The Man Who Lost His Language*, an outstanding book by Sheila Hale, about her husband John Hale's stroke. John was a famous historian, knighted for his contributions to British scholarship, Chair of the Trustees of the National Gallery and author of brilliant books, including *The Civilization of Europe in the Renaissance*. John's stroke-induced aphasia in effect transformed him back into the languageless infant he was at the beginning of his life. The tragic story (and indictment of the British National Health Service) is nevertheless a brilliant, touching, insightful account of this terrible deficit.

People with aphasia manifest one of a few different kinds of disorders. They can have problems with understanding what they hear (auditory comprehension) or difficulties communicating intelligibly, find reading and writing challenging or struggle to express themselves verbally. Aphasia is most commonly a result of extensive damage to the left hemisphere, the side of the brain long known to be implicated heavily in language and other tasks. Since the left hemisphere is, however, not exclusively dedicated to language, this means that aphasia will never affect language only. Some approaches to aphasia in the past have seen it largely in anatomical terms, localising it to what were considered once to be language specific regions of the brain, such as Broca's and Wernicke's areas. And indeed it is correct that there are specific kinds of aphasia associated with these general regions in the brain, as we have seen. Broca's aphasia, also known as 'motor aphasia' or 'expressive aphasia', is characterised by relatively solid comprehension, but difficulty in speaking. Significantly, however, people with this form of aphasia also usually have paralysis, or at least weakness, in their right appendages – the arm, the leg, or both.

Another well-known type of aphasia is Wernicke's aphasia, also known as 'receptive aphasia' or 'sensory aphasia'. In this form of aphasia, the subject is able to speak fluently but unable to understand what is said to them. Moreover, their 'fluent' speech often is chock-a-block with abnormalities, such as nonsense words that, while fitting the sound patterns of their native language, mean nothing and are otherwise not words at all. There are other forms of aphasia. This is, of course, unsurprising, since there are many aspects of language and many ways in which one might imagine it breaking down, especially if it is an artefact.

In an article in the journal *Aphasiology*, Edward Gibson and his co-authors offer an instructive analysis of strategies for overcoming aphasic language comprehension problems developed by the aphasics themselves.[6] What this team discovered is reminiscent of a G_1 language. The aphasics in Gibson's study use contextual clues more extensively than non-aphasic subjects in order to interpret what is spoken to them. *Homo erectus*, in my model at least, would have produced utterances that were highly ambiguous or vague or both, depending on his or her interlocutor's ability to link speech to context for an interpretation more or less in the ballpark of what was intended.* *Homo erectus* might, therefore, have used a strategy similar to aphasics for interpreting sentences. But that is what all people do. *All* utterances are interpreted at least in part based on knowledge of culture, the context and the world.

A discussion of language and the brain cannot end, however, before discussing another set of cognitive disorders, collectively known as autistic spectrum disorder. ASD reveals the importance of the society to language and the role of conversation as the apex of linguistic experience. Therefore, we must take special care in our discussion here. Richard Griffin and Daniel C. Dennett of Tufts University get at what seems to be the general thread running through many autism cases, namely that sufferers of autism share a 'pervasive bias to attend toward local rather than global features'. This is sometimes referred to as 'weak central coherence' and means that sufferers struggle to grasp an entire social situation in context.[7]

The first two people I can remember with ASD were friends of mine. One was a fellow redhead, brother of a grammar-school buddy. The other was a second cousin.

In the case of my friend's brother, I did not even know of his existence until I visited my friend's house and learned that he had a brother visiting from another school somewhere. He didn't seem old enough to be in college, so I couldn't really understand why he wouldn't be in school with us. As we were introduced, though, he was quiet and looked around erratically. I was about twelve years old. He would have been perhaps fourteen. He wouldn't answer me, so I asked my friend, 'Hey, what's wrong with your brother? Is he simple-minded?' My friend laughed. 'No, he is not simple. He's pretty smart. Wanna see?' I immediately replied, 'Yes!' though I was curious as to how my friend's brother would demonstrate his intelligence.

This was answered quickly. My friend took the annual calendar off of his bedroom wall and handed it to me. He suggested that I pick any

---

*This is reminiscent of what linguist Derek Bickerton refers to as 'protolanguage'. As I have remarked many times, however, I reject Bickerton's term here because it suggests (at least to me) that *erectus*'s language was known not to be a fully human language, whereas I consider *erectus*'s language fully developed, not merely a precursor to modern language.

month and ask his brother what day of the week any date was. 'What for?' 'Just ask him, you'll see.' So I thumbed through the calendar and picked a date earlier in the year. I think I asked, 'What is the day of January 21, 1963?' He immediately and correctly responded with the day of the week. I ran through about twenty dates all over the year. Then I asked him about other years. He never hesitated; he never erred. I was ready to acknowledge his intelligence as far exceeding my own. My friend's brother was giggling and giggled harder with every new date I asked him. In my twelve years, I had never heard of anything like that, much less witnessed such an example. 'How the hell's he do that?' I demanded. 'No idea,' replied my friend. He told his brother, 'Great job!'

I learned through this – though only later in life as I continued to think about what I had seen – that my friend's brother had at least two unusual characteristics: a seemingly deficient social ability and a brain that could do some things better than I had ever seen done. Yet the social difficulty meant that he struggled to carry on a conversation. My friend was particularly solicitous of his brother. 'He's done now,' he announced. 'Let's go.' And so we did.

My next encounter with what, in retrospect, was most probably also ASD was with a second-cousin I used to see daily at school. Other children teased him mercilessly. Once he yelled back at a tormentor with a nearly inarticulate, 'No, goddamn it!' and started hitting them with both hands. They initially started laughing. Then they realised he was really hurting them, so they apologised. My cousin stopped and walked away without another word. Neither I nor any of my friends at that time had a concept of neurodiversity. Anyone who didn't behave exactly as we expected we considered 'handicapped'.

So if I thought about my cousin, I felt pity for him. Then, one day, in an assembly of the entire junior high school that we both attended, the principal began to announce awards. Eventually he worked his way down the list to the two main awards of the year, scholarship and citizenship. When he said, 'For the first time both awards go to the same person,' I wasn't surprised. I expected them to go to a good friend of mine. But no. The name called, to my amazement, was my cousin's. For the first time in my life I realised that I was no judge of character. I teared up because I knew all the grief that my cousin put up with every day at school and he won citizenship! Conversations were just excruciatingly difficult for him. When he did talk, he became emotional and inarticulate very quickly. But he went to school. He ate alone. And then he impressed everyone with his intelligence and kindness.

There is, in fact, no single disease or aetiology that corresponds to what the general public calls 'autism', only a set of symptoms that professionals have decided to group under the general label of autistic spectrum disorder. Not everyone with this disorder behaves the same, as in the two different examples from my childhood. The symptoms of ASD include the following, broken into a couple of distinct areas:*

Communication: Autism sufferers struggle with communication and interactions with other people. People with this condition abnormally engage in routines or repetitive behaviours, sometimes called stereotyped behaviours. Autistics do not respond to their names by twelve months of age and cannot easily explain what they want. Generally those with autism do not follow directions. Sometimes they seem to hear, but not at other times. In general those with autism neither point nor wave 'bye-bye'.

Social Behaviour: People with autism generally don't smile when smiled at, make poor eye contact and appear to prefer to play alone. When fetching items or picking things up, they generally bring things for themselves only. Autism sufferers are independent for their age, yet give the impression of being in their 'own world'. They act as if they tune people out and are uninterested in other children, failing to point out interesting objects by fourteen months of age, a normal time frame for that development, and don't like to play 'peek-a-boo'. Sufferers do not try to attract their parents' attention.

Stereotyped Behaviour: Affected individuals get 'stuck' doing the same things over and over and can't move on, showing unusual attachments to toys, objects, or routines (always holding a string or having to put on socks before pants), and spend a lot of time lining things up or putting things in a certain order. They repeat words or phrases; sometimes called echolalia.

Other Behaviour: Children with autism may not play 'make believe'

---

*Paraphrased from the National Institutes of Health website www.nichd.nih.gov/health/topics/autism/conditioninfo/Pages/symptoms.aspx

or pretend by eighteen months of age. They have odd movement patterns and do not know how to play with toys appropriately, in spite of being attached strongly to them. Autism sufferers do some things 'early' compared to other children, such as walking on their toes, but do not like to climb on things such as stairs. These children don't imitate silly faces and seem to stare at nothing or wander around with no purpose. They throw intense or violent tantrums. Autistics may be overly active, uncooperative, or resistant and are overly sensitive to noise. The affected also prefer not to be swung or bounced on their parent's knee, etc.

A variety of causes have been proposed for ASD, including bad socialisation, bad parenting, bad genes, neurological developmental problems, excess testosterone and 'mind-blindness' (the purported lack of a theory of mind, that is a failure to recognise that others have a mind like the subject does). Research often produces messy results, though researchers often prefer elegant results. The idea that ASD is caused by 'mind-blindness' is a neat generalisation, but just doesn't seem to get an account of the spectrum of autistic disorders. None of these do. Rather, there is a sense in which they each play a role to a greater or lesser degree. As Dr Helen Tager-Flusberg, one of the world's leading autism researchers, puts it, 'Autism is a complex and heterogeneous disorder that should not be reduced to a single underlying cognitive impairment.'[8]

My own peculiar perspective on ASD as an outsider to the research and lacking the clinical expertise of real professionals, is that it is a breakdown of the ability to build components of the dark matter of the mind, structured cultural knowledge that underlies the development of the psychology of each individual as a cultural being. In other words, I agree with the general assessment in the literature that ASD is at root a social problem. But it is certainly a very unusual type of social problem, one which affects the ability to properly understand the intentions and thinking of others or to structure understanding about others from a background of cultural knowledge.

The symptoms of ASD are consistent with the idea that sufferers of the disorder are unable to incorporate other people's prioritisation or 'ranking' of values, perspectives, intentions, social roles and knowledge structures into the construction of their psyche. Thus they struggle to grasp the social implications of their apperceptions. But they show even more abnormalities, including certain bodily experiences. One possibility is that some of the physical aversions and preferences feed similar experiences in their minds. And ASD sufferers intensely dislike intrusions into their mental lives. Things such as sounds, touching, play, disorder that they cannot fix and phatic language distress sufferers. But, of course, these are crucial to constructing the articulated unconscious that is vital for one's appreciation for and place in the culture of our community.

Phatic language difficulties are particularly revealing, and include things like leave-takings ('Goodbye', 'Adios', 'Hasta la vista, baby' and 'Bye-bye'), greetings ('What's up!', 'Dude, what is it?', 'Hello!'), gratitude expressions ('Thanks', 'Oh wow, I can never repay you!') and so on. Phatic language has long been analysed by linguists and anthropologists as a form of 'grooming', that is a recognition of the other as a person you value, however superficially. Interestingly, although English – which has phatic language – lacks grooming as a regular cultural ritual across most of its communities, Pirahã, an Amazonian language, has daily grooming, in which men and women sit in lines and look for lice in each other's hair (or simply stroke each other's hair), yet it lacks phatic vocabulary. In other words, perhaps all cultures develop mechanisms, linguistic or otherwise, to demonstrate belonging and care to one another as members of the same community. ASD sufferers often seem to lack this capacity of mutual 'grooming' behaviours that establish that two or more individuals 'belong' together and accept one another.

Put in other ways, the ASD sufferer is often unable to build a theory either of their enveloping culture or of the intentions of those they interact with. Equally significantly, they are unable to develop a full range of cultural knowledge, values and appreciation of social roles in order to construct a social identity.

The isolation and dysfunctionality of ASD sufferers might be argued to arise from, in part, their inability to converse normally with others, itself resulting from a social inability to feel 'mutual belonging' and share 'mutual ideas'. To converse normally requires and helps to construct a reasonable command of the grammar of the language in which the conversation is taking place, an understanding of the context of

the conversation, the purpose of the conversation, the mental state of
the interlocutor(s), cultural background knowledge and general world
knowledge. These different abilities and forms of knowledge all boil
down to recognising culture and working to belong to one. These seem
to be the major ethnolinguistic difficulties faced by people with ASD.

Not only is conversation the apex of language, but engaging in
conversations has been shown by many researchers to build cultural
connections and knowledge while actually constructing the very
grammar that is necessary for the conversation. In other words, as we
have seen many times, language is synergistic, two or more components of or applications of language mutually enabling one another. In
this sense, the very existence of ASD underscores the importance of the
language–culture connection.

This discussion of the failure of sufferers of autism to construct
either a cultural role or robust cultural understanding raises another
proposal that has become popular among some in the discussion of
language evolution. This is what is referred to as 'niche construction'
theory and is the idea that humans fuse part of the environment, say
infancy and conversation, to create particular bio-cultural and psychological niches that enrich cognitive and linguistic development, such
that people are able to construct larger and larger niches. Perhaps, one
might reason, niche construction is the key to ASD – the child fails
to construct the proper 'niche'. In a sense this is correct. But another
special theory is unnecessary. ASD, so far as it relates to the connection between culture and the mind, is fully accounted for by what has
already been proposed, without need for further theories.

Niche theory is an elaborate mechanism to explain the development of the child and the species by focusing on the construction of
relationships of individuals via conversational interaction. There is
much to commend the work on niche construction. Yet at the same
time there are three problems that lead me to believe that it isn't very
helpful to either the modelling of individual human language development or maturation or to the understanding of human language
evolution. First, much of niche construction is already accounted for
by what psychologists refer to as 'attachment theory', which aims to
determine if there are principles that govern the growth in relationship
between infants and caregivers and, if so, what those principles are and

whether these principles are the same across cultures. (It appears that they might not be.)

The second reason one might not take niche theory to be an integral component of the story of language is that it is largely a metaphor for fairly well-understood phases of developmental psychology. And, finally, this is a theory that makes the wrong predictions about human language evolution, concluding that human language appeared only about 100,000 years ago, a conclusion claimed here to be incorrect.

This matter arises in various aspects of the invention of language, such as in the fact that language is 'just good enough'. Far from being a perfect biological system of any sort, language just gets by, often failing to communicate as well as one might imagine, its hearers employing general facts of the environmental context and world knowledge to interpret what is being said. And this echoes the strategies of aphasics and *Homo erectus*. Context and general knowledge are crucial to figure out the meanings of what people are hearing and in order to understand how to respond in the course of the conversation. For ASD sufferers, however, language is not even 'good enough'. It is deficient, because for these people their language is unable to link up with the social knowledge crucial to the major function of language, namely communication.

A startling conclusion emerges from deficits affecting language: *There are no language-only hereditary disorders.* And the reason for that is predicted by the theory of language evolution here – namely that there could not be such a deficit because there is no language-specific part of the brain. Language is an invention. The brain is no more specialised for language than for toolmaking, though over time both have affected the development of the brain in general ways that make it more supportive of these tasks.

Language disorders are a window into the human brain and its preparedness for language. But language is more than the brain. It is a function of the entire body, including those components from the lungs to the mouth that make oral speech possible. Although we know that language and speech are distinct, and that there are various kinds of 'speech' – visual speech, sign languages and vocal speech – the primary form of speech in all the world's languages is oral. So the question we must answer now that we understand the brain a bit better is how did evolution prepare us to vocalise our languages.

8

Talking with Tongues

... if the play makes the public aware that there are such people as phoneticians, and that they are among the most important people in England at present, it will serve its turn.

George Bernard Shaw, Preface to *Pygmalion*

In 1964 my eighth-grade school marching band won a local competition in the Imperial Valley of Southern California. I played baritone horn and was an enthusiastic member. And I knew that, by winning this contest, we would now be able to go to the higher regional competition in Los Angeles, about 130 miles (210 kilometres) to the northwest of our little town of Holtville, near the Mexican border.

Our director wanted to expose the band to some higher culture while in the LA area, so he petitioned the school board to allow us to attend a showing of Mozart's opera *Don Giovanni*. The school board said no. Too risqué for junior high schoolers. Instead, we were allowed to attend the band director's second choice, a showing at the Egyptian Theater in Hollywood of *My Fair Lady*, starring Rex Harrison and Audrey Hepburn. The band instructor prepared us by talking about the play by George Bernard Shaw, the source for the movie.

This film eventually played a role in my decision to become a linguist, as it revolved around the transformative power of human speech, told from the perspective of Henry Higgins and his reluctant pupil Eliza Doolittle. What is this thing called speech that all humans possess, that George Bernard Shaw believed to be the key to success in life? In *The Kingdom of Speech* Tom Wolfe claims that speech is the most important invention in the history of the world. It not only enables us to speak to one another but also immediately classifies by economic class, age group and educational attainment. If *erectus* were around today, would people consider them brutish because of the way they spoke, even if one could so dress them as to pass them off as a peculiar-looking modern human?

Although communication is ancient, human speech is evolutionarily recent. Cognitive scientist and phonetician Philip Lieberman claims that the speech apparatus of modern *Homo sapiens* is only about 50,000 years old, so recent that even earlier *Homo sapiens* could not speak as we do today.[1] This is not to be confused, though, with the 50,000 year date proposed by other authors on the appearance of language. Speech followed language. Therefore, Lieberman's 50,000 year date would be, if correct, evidence *against* the idea that language proper appeared suddenly 50,000 years ago. If *erectus* did indeed invent symbols and begin humanity's upward trek through the progression of signs to language, enhanced speech would have come later. It is expected that the first languages would have been inferior to our present languages. No invention begins at the top. All human inventions get better over time. And yet this does not mean that *erectus* spoke a subhuman language. What it does mean, however, is that they lacked fully modern speech, for physiological reasons, and that their information flow was slower – they didn't have as much to talk about as we do today, nor do they seem to have had sufficient brain power to process and produce information as quickly as modern *sapiens*. *Erectus*'s physiological shortcoming was overcome by gradual biological evolution. The development of information processing and enhanced grammatical ability results from cultural evolution. Both biological and cultural evolution over 60,000 subsequent generations of humans improved our linguistic abilities dramatically.

In a 2016 paper in the publication of the American Association for the Advancement of Science, Tecumseh Fitch and his colleagues argue in effect that Lieberman is mistaken in his view of the evolution of the human vocal tract. They claim instead that the vocal apparatus is much older than the fifty millennia proposed by Lieberman – so old, in fact, that it is found in macaque monkeys.[2] While the study by Fitch and colleagues is intriguing, there are two reasons why it is not particularly useful in understanding language evolution. First, most of the extreme tongue positions that they claim to be similar between macaques and

humans come from macaque's yawning. The assumption of Fitch and
his co-authors seems to be that, if they can get macaques to place their
tongues in the right positions to produce certain human vowels while
yawning, the macaques could repeat this if they were speaking. This is a
doubtful assumption, though, because yawning is not nearly so effort-
less a gesture as the production of a back vowel (similar in the shape
of the tongue for yawning) is for humans. The tongue is retracted in
effortful ways, and it is doubtful speech sounds would evolve out of a
yawning vocal tract shape. Another problem with this study is that the
authors compare the phonetics of the macaque against archival human
phonetics. But the authors should have retested human phonetic properties using the very same methods they used on macaques, in order to
compare them more equally.[3] Finally and most importantly, is the fact
that language does not require speech as we know it. Languages can
be whistled, hummed, or spoken with a single vowel with or without a
consonant. It is the confluence of culture and the *Homo* brain that gives
us language. Our modern speech is a nice, functional add-on.

On the surface of it, human speech is simple. Vowels and consonants are created along the same principles as the notes formed by
wind blowing through a clarinet. The root of both is basic physics. Air
flows up from the lungs and out the mouth and is modified as it passes
through either the tube of the clarinet or the tube of the human vocal
apparatus. In the case of the clarinet the airflow is transformed by keys
and a reed that alter it so that it can make the sublime sounds Benny
Goodman produced, or the squeaking and squawking of a beginner.
Before reaching the mouth, the flow of air is transformed into speech
by the larynx, the tongue, the teeth and the different shapes and movements of all the stuff in our throats, noses and mouths that lie above
the larynx.

But speech is more complex than a mere wind-tube effect. That is
because the tube of human speech is controlled by a complex respiratory physiology directed by an even more complex brain. The creation
of speech requires precise control of more than one hundred muscles
of the larynx, the respiratory muscles, the diaphragm and the muscles
between our ribs – our 'intercostal' muscles – and muscles of our mouth
and face – our orofacial muscles. The muscle movements required of all
these parts during speech is mind-bogglingly complex. The ability to

make these movements required that evolution change the structures of the brain and the physiology of the human respiratory apparatus. On the other hand, none of these subsequent adaptations was required for language. They all simply made speech the highly efficient form of transmitting language that we know it as today. Still, it is unlikely that any *erectus* woman could have played Eliza. Her appearance would never have fooled anyone.

There are three basic parts to human speech capabilities that evolution needed to provide us with to enable us to talk and sing as humans do today. These are the lower respiratory tract, which includes the lungs, heart, diaphragm and intercostal muscles, the upper respiratory tract, which includes the larynx, the pharynx, the nasopharynx, the oropharynx, the tongue, the roof of the mouth, the palate, the lips, the teeth and, most importantly by far, the brain.

The average human produces 135–185 words per minute. Two things about this are deeply impressive. First, it is amazing that humans can talk that fast and consider it normal. Second, it is nearly incredible that people can understand anyone speaking that fast. But, of course, humans do both of these, producing and perceiving speech, without the slightest effort when they are healthy. These are the two sides to speech – production (speaking or signing) and perception (hearing with understanding). To grasp how speech production and perception evolved one has to know not only how the upper and lower respiratory tracts evolved, but also how the brain is able to control the physical components of speech so well and so quickly.

To tell the story of speech, we need to look at the vocal apparatus and the evidence for speech capabilities across the various species of *Homo*. It is important to have a clear idea of how sounds are made, how sounds are perceived and how the brain is able to manage all of this. But prior to that, it is essential to comprehend the state of human speech today. How does speech work with modern *Homo sapiens*? Knowing the answer to such questions can make it possible to judge how effective the speech of other *Homo* species would have been relative to *sapiens'*, and if they were, in fact, capable of speech.

Speech comes out of mouths, travels through the air and enters the ears of hearers, to be interpreted by their brains. Each of the three steps in the creation and transmission and understanding of speech has an entire subfield of phonetics, the science of sounds, dedicated to it. The creation of sounds is the domain of the field of 'articulatory phonetics'. The transmission of sounds through the air is 'acoustic phonetics'. And the hearing and interpretation of sounds is 'auditory phonetics'. But one also encounters names of subfields that are concerned with other types of function. There are studies of the physics and mechanics of speech perception and speech production. These different studies are often grouped together under the name 'experimental phonetics'. It isn't necessary to understand all of these to understand the evolution of speech but a wee bit of understanding of them would be helpful.

The larynx is vital to the understanding of the language of *Homo* species, as it enables humans not only to pronounce human speech sounds, but also to have intonation and use pitch to indicate what aspect of an utterance is new, what is old, what is particularly important, whether people are asking a question, or making a statement. The larynx is where the airflow from the lungs is manipulated in order to produce phonation, the confluence of energy, muscles and airflow required to produce the sounds of human speech.

The larynx is a small transducer that sits atop the trachea, with a top called the epiglottis that can flip closed to keep food or liquid from entering through the larynx and into the lungs, potentially causing great harm. Figure 21 shows just a glimpse of its complexity.*

One thing that every researcher into the evolution of speech agrees upon is the idea that our speech production evolved in tandem with our speech perception. Or, as Crelin puts it in his pioneering work, 'there tends to be a precise match between the broadcast bandwidth and the tuning of perceptual acuity'. Or 'the possession of articulate speech therefore implies that both production and perception are attuned to each other, so that parameters carrying the bulk of the

---

*For those interested in the history of studies of human speech, these go back for centuries. But the modern investigation of the physiology and anatomy of human speech is perhaps best exemplified in a book by Edmund S. Crelin of the Yale University School of Medicine, entitled *The Human Vocal Tract: Anatomy, Function, Development, and Evolution* (New York: Vantage Press, 1987). It contains hundreds of drawings and photographs not only of the modern human vocal apparatus but also of the relevant sections of fossils of early humans, as well as technical discussions of each.

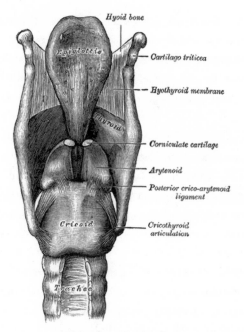

Figure 21: **The larynx**

speech information are optimized in both production and perception'. In other words, the ears and the mouth work well together because they have evolved together for several million years.

Speech begins with air, which can create human sounds when it flows into the mouth or when it is expelled from the mouth. The former types of sounds are called 'ingressive' sounds and the latter are 'egressive' sounds. English and other European languages use egressive sounds exclusively in normal speech. Ingressive sounds in these better-known languages are rare, usually found only in interjections, such as the sound of 'huh' when the air is sucked in. The place where the air begins in a sound is called the 'initiator'. In all of the speech sounds of English, the lungs are the initiator. Thus one says that all sounds of English are 'pulmonic' sounds. But there are two other major air initiators that many languages of the world use, the glottis (the opening in the larynx, for glottalic sounds) and the tongue (for lingual sounds). These are also not sounds of English.

To quote from my *Language: The Cultural Tool*:

[In] Tzeltal, Ch'ol and others, so-called 'glottalised' sounds – implosives and ejectives – are common.

When I began my linguistic career, in the mid-1970s, I went to live for several months among the Tzeltales of Chiapas, Mexico. One of my favourite phrases was *c'uxc'ajc'al* 'it's hot outside', which contains three glottalised consonants (indicated in Tzeltal orthography by the apostrophe). To make these sounds, the glottis, the space between the two vocal cords in the larynx, must be closed, cutting off air from the lungs. If the entire larynx is then forced up at the same time that the lungs or tongue cut off the flow of air out of the mouth, then pressure is created. When the tongue or lips then release the air out of the mouth, an explosive-like sound is produced. This type of sound, seen in the Tzeltal phrase above, is called an 'ejective'. We could also produce the opposite of an ejective, known as an 'implosive' sound. To make an implosive, the larynx moves down instead of up, but everything else remains the same as for an ejective. This downward motion of the larynx will produce an implosive – caused by air suddenly rushing into the mouth. We do not have anything like these sounds in English.

I remember practising ejectives and implosives constantly for several days, since the Tzeltales I worked with use them both. They are interesting sounds – not only are they fun sounds, but they extend the range of human speech sounds beyond the strictly lung-produced sounds in European languages.

The glottis can be used to modify sounds in other ways. Again, from *Language*:

A different type of glottalised sound worth mentioning is produced by nearly, but not quite, closing the glottis and allowing lung air to barely flow out. This effect is what linguists call 'creaky voice'. People often produce creaky voice involuntarily in the mornings after first arising, especially if their vocal cords are strained through yelling, drinking, or smoking. But in some languages, creaky voice sounds function as regular vowels.

Some glottalised sounds are known as clicks. These are created by

THE INTERNATIONAL PHONETIC ALPHABET (revised to 2015)

CONSONANTS (PULMONIC)   © 2015 IPA

	Bilabial	Labiodental	Dental	Alveolar	Postalveolar	Retroflex	Palatal	Velar	Uvular	Pharyngeal	Glottal
Plosive	p b			t d		ʈ ɖ	c ɟ	k g	q ɢ		ʔ
Nasal	m	ɱ		n		ɳ	ɲ	ŋ	ɴ		
Trill	ʙ			r					ʀ		
Tap or Flap		ⱱ		ɾ		ɽ					
Fricative	ɸ β	f v	θ ð	s z	ʃ ʒ	ʂ ʐ	ç ʝ	x ɣ	χ ʁ	ħ ʕ	h ɦ
Lateral fricative				ɬ ɮ							
Approximant		ʋ		ɹ		ɻ	j	ɰ			
Lateral approximant				l		ɭ	ʎ	ʟ			

Symbols to the right in a cell are voiced, to the left are voiceless. Shaded areas denote articulations judged impossible.

CONSONANTS (NON-PULMONIC)

Clicks		Voiced implosives		Ejectives	
ʘ	Bilabial	ɓ	Bilabial	ʼ	Examples:
ǀ	Dental	ɗ	Dental/alveolar	pʼ	Bilabial
ǃ	(Post)alveolar	ʄ	Palatal	tʼ	Dental/alveolar
ǂ	Palatoalveolar	ɠ	Velar	kʼ	Velar
ǁ	Alveolar lateral	ʛ	Uvular	sʼ	Alveolar fricative

OTHER SYMBOLS

ʍ Voiceless labial-velar fricative   ɕ ʑ Alveolo-palatal fricatives
w Voiced labial-velar approximant   ɺ Voiced alveolar lateral flap
ɥ Voiced labial-palatal approximant   ɧ Simultaneous ʃ and x
ʜ Voiceless epiglottal fricative
ʢ Voiced epiglottal fricative   Affricates and double articulations
ʡ Epiglottal plosive   can be represented by two symbols
                                         joined by a tie bar if necessary.   t͡s k͡p

DIACRITICS   Some diacritics may be placed above a symbol with a descender, e.g. ŋ̊

̥ Voiceless	n̥ d̥	̤ Breathy voiced	b̤ a̤	̪ Dental	t̪ d̪
̬ Voiced	s̬ t̬	̰ Creaky voiced	b̰ a̰	̺ Apical	t̺ d̺
ʰ Aspirated	tʰ dʰ	̼ Linguolabial	t̼ d̼	̻ Laminal	t̻ d̻
̹ More rounded	ɔ̹	ʷ Labialized	tʷ dʷ	̃ Nasalized	ẽ
̜ Less rounded	ɔ̜	ʲ Palatalized	tʲ dʲ	ⁿ Nasal release	dⁿ
̟ Advanced	u̟	ˠ Velarized	tˠ dˠ	ˡ Lateral release	dˡ
̠ Retracted	e̠	ˤ Pharyngealized	tˤ dˤ	̚ No audible release	d̚
̈ Centralized	ë	̴ Velarized or pharyngealized	ɫ		
̽ Mid-centralized	e̽	̝ Raised	e̝ (ɹ̝ = voiced alveolar fricative)		
̩ Syllabic	n̩	̞ Lowered	e̞ (β̞ = voiced bilabial approximant)		
̯ Non-syllabic	e̯	̘ Advanced Tongue Root	e̘		
˞ Rhoticity	ɚ a˞	̙ Retracted Tongue Root	e̙		

VOWELS

Where symbols appear in pairs, the one to the right represents a rounded vowel.

SUPRASEGMENTALS

ˈ Primary stress   ˌfoʊnəˈtɪʃən
ˌ Secondary stress
ː Long   eː
ˑ Half-long   eˑ
̆ Extra-short   ĕ
| Minor (foot) group
‖ Major (intonation) group
. Syllable break   ɹi.ækt
‿ Linking (absence of a break)

TONES AND WORD ACCENTS

LEVEL			CONTOUR		
e̋ or ˥	Extra high		ě or ˇ	Rising	
é or ˦	High		ê	Falling	
ē or ˧	Mid		e᷄	High rising	
è or ˨	Low		e᷅	Low rising	
ȅ or ˩	Extra low		e᷈	Rising-falling	
↓ Downstep			↗ Global rise		
↑ Upstep			↘ Global fall		

Figure 22: The International Phonetic Alphabet

using the tongue to block the flow of air into or out of the mouth, while pressure builds up behind the glottis. As with other sounds using the lungs or the glottis, lingual sounds can also be egressive or ingressive, produced by closing the airflow off with the tip of the tongue while building pressure inward or outward with the back of the tongue. Clicks are found in a vary small number of languages, all in Africa and almost all languages of the Bantu family. I remember first hearing clicks in Miriam Makeba's 'click song'. Makeba's native language was Xhosa, a Bantu language.

A list of all the consonants that are produced with lung air are given in the portion of the International Phonetic Alphabet shown in Figure 22.

Consonants are different from vowels in several ways. Unlike vowels, consonants impede (rather than merely shaping) the flow of air as it comes out of the mouth. The International Phonetic Alphabet (IPA) chart is recognised by all scientists as the accepted way of representing human speech sounds. The columns of the chart are 'modes' of pronunciation. These modes include allowing the air to flow out of the nose, which produces nasal sounds like [m], [n] and [ɲ]. Other modes are 'occlusives' or 'stops' (air is completely blocked as it flows through the mouth), sounds such as [d], [t], [k], or [g]. And there are 'fricatives', where the airflow is not completely stopped, but it is impeded sufficiently to cause friction, turbulent or sibilant sounds, such as [s], [f] and [h].

The rows in the IPA chart are places of articulation. The chart starts on the left with sounds produced near the front of the mouth and moves further back in the throat. The sounds [m] and [b] are 'bilabials'. They are produced by blocking the flow of air at the lips, both upper and lower lips coming together to completely block the flow of air. The sound [f] is a bit further back. It is produced by the lower lip touching the upper teeth and only partially impeding rather than completely blocking the flow of air. Then we get to the sounds like [n], [t] and [d], where the tongue blocks the flow of air either just behind the teeth (as in Spanish) or at the small ridge (alveolar ridge) on the hard palate (roof of the mouth), not far behind the teeth (as in English).

We eventually get to the back of the mouth, where sounds like [k] and [g] are produced by the back of the tongue rising to close off air at the soft palate. In other languages, the sounds go back further. Arabic languages are known for their pharyngeal sounds, made by constricting the epiglottis or by retracting the tongue into the pharynx. The epiglottis is a piece of stretchy cartilage that comes down to cover the hole at the top of the larynx just in case food or liquid tries to get in. One should not talk with a full mouth; if the epiglottis is not at the ready this could be fatal. Humans, except human infants, are the only creatures that cannot eat and vocalise at the same time.

What is crucial in the IPA charts is that the segments they list almost completely exhaust all the sounds that are used for human languages anywhere in the world. The phonetic elements therein are all easy (at least with a bit of practice) for humans to make. But the basal ganglia do favour habits, and so once we have mastered our native set of phonemes, it can be hard to get the ganglia out of their rut to learn the articulatory habits necessary for the speech of other languages.

But consonants do not speech make. Humans also need vowels. To take one example, the vowels of my dialect of English, Southern California, are as shown in Figure 23.

Just like the consonant chart, the vowel chart in Figure 23 is 'iconic'. Its columns represent the front of the mouth to the back. The rows of the vowel chart indicate the relative height of the tongue when producing the vowel. The trapezoidal shape of the chart is intended to indicate, again iconically, that as the tongue lowers, the space in the mouth between the vowels shrinks.

California vowels, like all vowels, are target areas where the tongue raises or lowers to a specific region in the mouth. At the same time, as the tongue moves to the target area to raise or lower, the tongue muscles are tense or relaxed ('lax'). The lips can be rounded or flat. The tense vowel, [i] is the vowel of the word 'beet'. The lax vowel [ɪ], on the other hand, is the vowel heard in the word 'bit'. In other words 'bit' and 'beet' are identical except that the tongue muscles are tense in the 'beet' and relaxed in 'bit'. Another way of talking about 'lax' vs 'tense' vowels, one preferred by many linguists, is to refer to them as 'Advanced Tongue Root' (the tongue is thereby tensed by being pushed forward in the mouth and flexing) or 'Not Advanced Tongue Root' (the tongue is relaxed, its root further back in the mouth), usually written in the linguistics literature as [+ATR] or [−ATR].

The funny-looking vowel character [æ] is the vowel in my dialect

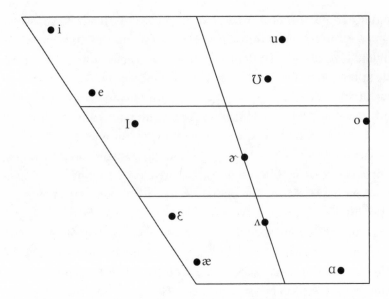

Figure 23: **Southern California English vowels**

for 'cat'. It is low, front and unrounded. But if moving up the chart and to the back of the mouth the sound [u] is reached, the vowel sound of the word 'boot'. This is a back, rounded vowel. 'Back' in this sense means that the back portion of the tongue is raised, rather than the front portion (the blade or tip) as in the unrounded vowel [i]. The lips form a round 'o' shape when producing [u]. Any vowel can be rounded. Thus, to make the French vowel [y], make the English vowel [i] while rounding the lips.

The point is that the various speech sounds available to all human languages are conceptually easy to understand. What is hard about them is not how to classify nor even to analyse them, but how to produce them. Humans can learn all the sounds they want when they are young and their basal ganglia are not in a rut. But as we get older, the ganglia are challenged to make new connections.

When I was enrolling for my first course in articulatory phonetics (in order to learn how to make all the sounds of speech used in all the languages of the world) at the University of Oklahoma in 1976, the teaching assistants for the course gave each student an individual interview in order to place them in sections according to phonetic 'talent' (or perceptual ability). I walked into the classroom used for this purpose and the first thing I was asked to do was to say the word 'hello' – but by sucking air into my lungs rather than expelling air outwards. 'Weird,' I thought. But I did it. Then I was asked to imitate a few words from Mayan languages with glottal ejectives. These are 'popping' sounds in which the air comes out of the mouth but originates above the lungs, with pressure built up by bringing the vocal folds together and then letting the air behind them 'eject' out of the mouth. And I tried imitating African click sounds. This course was going to be valuable to me, I knew, because I was preparing to go to the Amazon to conduct field research on a language that was still largely unknown to the outside world, Pirahã.

Again, every language in the world, from Armenian to Zapotec, uses the same inventory of articulatory movements and sounds. The reason for this is that the human auditory system co-evolved with the human articulatory system – that is, humans learned to hear best the sounds they are able to make. There are always outliers, of course, and there are still unexpected novelties to be discovered. In fact, I have personally discovered two sounds in the Amazon over the years (one in the Chapakuran language family, the other in Pirahã) not found in any other language of the world.

The field linguist needs to learn what sounds the human body can make and use for speech because she or he must be prepared to begin work from the very first minute they arrive at their field destination. They have to know what they are hearing in order to begin an analysis of the speech and language of the people they have gone to live with.

This brief introduction covers only part of one-third of the science of phonetics, articulatory phonetics. But what happens to speech sounds once they exit the mouth? How are people able to distinguish them? Hearers are not usually able to look into the mouth of the person speaking to them, so how can they tell whether she or he is making a [p] or a [t] or a [i] or a [a]?

This is the domain of acoustic phonetics. An immediate question regarding sound perception is why, if air is coming out of the mouth when one is talking, is it that only the consonants and vowels are heard, rather than the sound of air rushing out of the mouth? First, the larynx energises the air by the vibration of the vocal cords, or oscillation of other parts of the larynx. This changes the frequency of the sound and brings it within the perceptible range for humans because evolution has matched these frequencies to human ears. Second, the sounds of the air rushing out of the mouth have been tuned out by evolution, falling below the normal frequency ranges that the human auditory system can easily detect. That is a good thing. Otherwise people would sound like they're wheezing instead of talking.

The energising of the flow of air in speech by the larynx is known as phonation, which produces for each sound what is known as the 'fundamental frequency' of the sound. The fundamental frequency is the rate of vibration of the vocal folds during phonation and this varies according to the size, shape and bulk (fat) of the larynx. Small people will generally have higher voices, that is, higher fundamental frequencies, than larger people. Adults have deeper voices, lower fundamental frequencies, than children, men have lower voices than women and taller people often have deeper voices than shorter people.

The fundamental frequency, usually written as F_0, is one of the ways that people can recognise who is talking to them. We grow accustomed to others' frequency range. The varying frequency of the vibration of the vocal cords is also how people sing and how they control the relative pitches over syllables in tone languages, such as Mandarin or Pirahã, among hundreds of others where the tone on the syllable is as important to the meaning of the word as the consonants and vowels. This ability to control frequency is also vital in producing and perceiving the relative pitches of entire phrases and sentences, referred to as intonation. F_0 is also how some languages are whistled, using either the relative pitches on syllables or the inherent frequencies of individual speech sounds.

It may surprise no one to learn, however, that F_0 is not all there is. In addition to the fundamental frequency, as each speech sound is produced harmonic frequencies, or *formants*, are produced that are associated uniquely what that particular sound. These formants enable us to distinguish the different consonants and vowels of our native language. One does not directly hear the syllable [dad], for example. What is heard are the formants and their changes associated with these sounds.

A formant can be visualised by hitting a tuning fork that produces the note 'E' and placing it on the face of an acoustic guitar near the sound hole. If the guitar is tuned properly, the 'E' string of the same

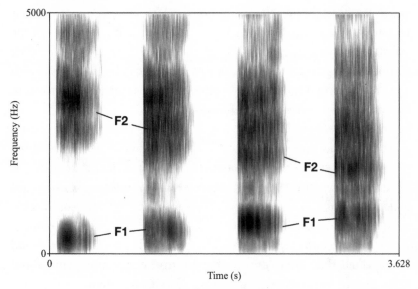

Figure 24: **Vowel spectrogram**

octave as the tuning fork will vibrate or resonate with the fork's vibrations. This resonance is responsible for the different harmonics or formants of each speech sound. These formants can be seen in a spectrogram, with each formant at a particular multiple of the fundamental frequency of the sound (Figure 24).

In this spectrogram of four vowels the fundamental frequency is visible at the bottom and dark bands are visible going up the columns. Each band, associated with a frequency on the left side of the spectrogram, is a harmonic resonance or formant of the relevant vowel. Going left to right across the bottom, the time elapsed in the production of the sounds is measured. The darkness of the bands indicates the relative loudness of the sound produced. It is the formants that are the 'fingerprints' of all speech sounds. Human ears have evolved to hear just these sounds, picking out the formants which reflect the physical composition of our vocal tract. The formants, from low to high frequency, are simply referred to as F_1, F_2, F_3 and so on. They are caused by effects of resonators such as the shape of the tongue, rounding of the lips and other aspects of the sound's articulation.

The formant frequencies of the vowels are seen in the spectrogram (given in hertz (Hz)). What is amazing is not that only that we hear these frequency distinctions between speech sounds, but that we do so without knowing what we are doing, even though we produce and perceive these formants so unerringly. It is the kind of tacit knowledge that often leads linguists to suppose that these abilities are congenital rather than learned. And certainly some aspects of this are inborn. Human mouths and ears are a matched set, thanks to natural selection.

There is too little understood about how sounds are interpreted physiologically by our ears and brains to support a detailed discussion of auditory phonetics, the physiology of hearing. But the acoustics and articulation of sounds is quite enough to prime a discussion of how these abilities evolved.

If it is correct to say that language preceded speech, then it would be expected that *Homo erectus*, having invented symbols and come up with a G_1 language, would still not have possessed top-of-the-line human speech capabilities. And they did not. Their larynges were more ape-like than human like. In fact, although *neanderthalensis* had relatively modern larynges, *erectus* lagged far behind.

The main differences between the *erectus* vocal apparatus and the *sapiens* apparatus were in the hyoid bone and pre-*Homo* vestiges, such as air sacs in the centre of the larynx. Tecumseh Fitch was one of the first biologists to point out the relevance of air sacs to human vocalisation. Their effect would have been to render many sounds emitted less clear than they are in *sapiens*. The evidence that they had air sacs is based on luck in finding fossils of *erectus* hyoid bones. The hyoid bone sits above the larynx and anchors it via tissue and muscle connections. By contracting and relaxing the muscles connecting the larynx to the hyoid bone, humans are able to raise and lower the larynx, altering the F_0 and other aspects of speech. In the hyoid bones of *erectus*, on the other hand, though not in any fossil *Homo* more recent than *erectus*, there are no places of attachment to anchor the hyoid. And these are not the only differences. So different are the vocal apparatuses of *erectus* and *sapiens* that Crelin concludes: 'I judge that the vocal tract was basically apelike.' Or, as others say:

Authors describe a hyoid bone body, without horns, attributed to *Homo erectus* from Castel di Guido (Rome, Italy), dated to about

400,000 years BP. The hyoid bone body shows the bar-shaped morphology characteristic of *Homo*, in contrast to the bulla-shaped body morphology of African apes and *Australopithecus*. Its measurements differ from those of the only known complete specimens from other extinct human species and early hominid (Kebara Neanderthal and *Australopithecus afarensis*) and from the mean values observed in modern humans. The almost total absence of muscular impressions on the body's ventral surface suggests a reduced capability for elevating this hyoid bone and modulating the length of the vocal tract in *Homo erectus*. The shield-shaped body, the probable small size of the greater horns and the radiographic image appear to be archaic characteristics; they reveal some similarities to non-humans and pre-human genera, suggesting that the morphological basis for human speech didn't arise in *Homo erectus*.[4]

There is no way that *erectus*, therefore, could have produced the same kind or quality of speech, in terms of ability to clearly discriminate the same range of speech sounds in perception or production, as modern humans. None of this means that *erectus* would have been incapable of language, however. *Erectus* had sufficient memory to retain a large number of symbols, at least in the thousands – after all, dogs can remember hundreds – and would have been able, in the use of context and culture, to disambiguate symbols that were insufficiently distinct in their formants, due to the lesser articulatory capacity of *erectus*. However, what is expected is that the new dependency on language would have created a Baldwin effect such that natural selection would have preferred *Homo* offspring with greater speech production and perception abilities, both in the vocal apparatus as well in the various control centres of the brain. Eventually, humans went from *erectus*'s low-quality speech to their current high-fidelity speech.

What size inventory of consonants and vowels, intonation and gestures does a language need to ensure that it has the right 'carrying capacity' for all the meanings it wants to communicate? Language can be thought of in many ways. One way to view it, though, is as matching up meanings to forms and knowledge in such a way that hearers can understand speakers.

If it were known with certainty that *Homo erectus* and *Homo neanderthalensis* were incapable of making the full range of sounds of anatomically modern humans, would this mean that they could not have had languages as rich as *sapiens*? It is hard to say. It is almost certain that *sapiens* are better at speech than *erectus* and other hominins that preceded *sapiens*. There are innumerable benefits and advantages to being the proud possessor of a modern speech apparatus. It makes speech easier to understand. But *sapiens*' souped-up vocal tract is not necessary to either speech or language. It is just very, very good to have. Like having a nice travel trailer and a powerful 4x4 pickup instead of a covered wagon pulled by two mules.

In fact, computers show that a language can work just fine with only two symbols, *o* and *1*. All computers communicate by means of those two symbols, current on – 1 – and current off, o. All the novels, treatises, PhD dissertations, love letters and so on in the history of the world can, with many deficiencies, such as the absence of gestures, intonation and information about salient aspects of sentences, be translated into sequences of o and 1. So, if *erectus* could have made just a few sounds, more or less consistently, they could be in the language game, right there with *sapiens*. This is why linguists recognise that language is distinct from speech. *Sapiens* quite possibly speak more clearly, with sounds that are easier to hear. But this only means, again, that *erectus* drove a Model T language. *Sapiens* drive the Tesla. But both the Model T and the Tesla are cars. The Model T is not a 'protocar'.

Though hard to reconstruct from fossil records, the human vocal tract, like the human brain, also evolved dramatically from earlier hominids to modern *sapiens*. But in order to tell this part of the story, it is necessary to back up a bit and talk about the sounds that modern humans use in their languages. This is the end point of evolution and the starting point of any discussion of modern human speech sounds.

The evolutionary questions that lie beneath the surface of all linguistics field research are 'How did humans come to make the range of sounds that are found in the languages of the world today?' and, next, 'What are these sounds?'

The sounds that the human vocal apparatus uses all are formed from the same ingredients.

The technical label for any speech sound in any language tells how that sound is articulated. The consonant [p] is known as a: 'voiceless bilabial occlusive (also called a "stop") with egressive lung air'. This long, but very helpful, description of the speech sound means that the [p] sound in the word 'spa', to take one example, is pronounced by relaxing the vocal cords so that they do not vibrate. The sound is therefore 'voiceless'. (The sound [b] is pronounced exactly like [p] except that in [b] the vocal folds – also called cords – are tensed and vibrating, rendering [b] a 'voiced' sound.) The phrase 'egressive lung air' means that the air is flowing out of the mouth or nose or both and that it originated with the lungs. This needs to be stated because not all speech sounds use lung air. The term 'occlusive', or 'stop', means that the airflow is blocked entirely, albeit momentarily. The term 'bilabial' refers to the action of the upper and lower lips together. In conjunction with the term 'occlusive', 'bilabial' means that the airflow was blocked entirely by the lips. If one pronounces the sounds in the hypothetical word [apa] while lightly holding the index finger on the 'Adam's apple' (which is actually your larynx), the vibration of the vocal cords can be felt to cease from the first [a] to the [p] and then start vibrating again on the second [a]. But if the same procedure is followed for the hypothetical word [aba] the vocal cords will continue to vibrate for each of [a], [b] and [a], that is for the entire duration of the word.

Though there are hundreds of sounds in the world's 7,000+ languages, they are all named and produced according to these procedures. And what is even more important, these few simple procedures, using parts of the body that evolved independently of language – the teeth, tongue, larynx, lungs and nasal cavity – are sufficient to say anything that can be said in any language on the planet. Very exciting stuff.

Humans can, of course, bypass speech altogether and communicate with sign languages or written languages. Human modes of communication, whether writing, sign languages or spoken languages, engage one or both of two distinct channels – the aural-oral and the manual-visual. In modern human languages, both channels are engaged from start to finish. This is essential in human language, where gestures, grammar and meaning are combined in every utterance humans make. There are other ways to manifest language, of course. Humans can communicate using coloured flags, smoke signals, Morse code, typed letters, chicken entrails and other visual means. But, funnily enough, no one expects to find a community that communicates exclusively by writing or smoke signals unless they have some sort of shared physical challenge or are all cooperating with some who do.

One question worth asking is whether there is anything special about human speech or whether it is just composed of easy-to-make noises.[5] Would other noises work as well for human speech?

As Philip Lieberman has pointed out, one alternative to human speech sounds is Morse code.[6] The fastest speed a code operator can achieve is about fifty words per minute. That is about 250 letters per minute. Operators working this quickly, however, need to rest frequently and can barely remember what they have transcribed. But a hungover college student can easily follow a lecture given at the rate of 150 words per minute! We can produce speech sounds at roughly twenty-five per second.

Speech works also by structuring the sounds we make. The main structure in the speech stream is the syllable. Syllables are used to organise phonemes into groups that follow a few highly specific patterns across the world's languages.[7] The most common patterns are things like Consonant (C) + Vowel (V); C+C+V; C+V+C; C+C+C+V+C+C+C and so on (with three consonants on either side of the vowel pushing the upper limits of the largest syllables observed in the world's languages). English provides an example of complex syllable structure, seen in words like *strength, s-t-r-e-n-g-th,* which illustrates the pattern C+C+C+V+C+C+C (with 'th' representing a single sound). But what I find interesting is that in the majority of languages C+V is either the only syllable or by far the most common. With the organisational and mnemonic help of syllables, our neural evolution plus our contingency judgements – based on significant exposure to our native language – we are able to parse our speech sounds and words far faster than other sounds.

Suppose you want to say, 'Piss off, mate!' How do you get those sounds out of your mouth on their way to someone's ear? There are three syllables, five consonants and three vowels in these three words, based on the actual spoken words rather than the written words using the English alphabet. The sounds are, technically, [pʰ], [I], [s], [ɔ], [f], [m], [eⁱ] and [t]. The syllables are [pʰIs], [f] and [meⁱt] and so, unusually in English, each word of this insult is also a syllable.

Sign languages also have much to teach us about our neural cognitive-cerebral platform. Native users of sign languages can communicate as quickly and effectively as speakers using the vocal apparatus. So our brain development cannot be connected to speech sounds so tightly as to render all other modalities or channels of speech unavailable. It seems unlikely that every human being comes equipped by evolution with separate neuronal networks, one for sign languages and another for spoken languages. It is more parsimonious to assume instead that our brains are equipped to process signals of different modalities and that our hands and mouths provide the easiest ones. Sign languages, by the way, also show evidence for syllable-like groupings of gestures, so we know that we are predisposed to such groupings, in the sense that our minds quickly latch on to syllabic groupings as ways of better processing parts of signs. Regardless of other modalities, though, the fact remains that vocal speech is the channel exclusively used by the vast majority of people. And this is interesting, because in this fact we do see evidence that evolution has altered human physiology for speech.

Human infants begin life much as other primates vocally. The child's vocal tract anatomy above their larynx (the supralaryngeal vocal tract or SVT) is very much like the anatomy of the corresponding tract in chimps. When human newborns breathe, their larynx rises to lock into the passage leading to the nose (the nasopharyngeal passage). This seals off the trachea from the flow of mother's milk or other things in the newborn's mouth. Thus human babies can eat and breathe without choking, just like a chimp.

Adults lose this advantage. As they mature, their vocal tract elongates. Their mouths get shorter while the pharynx (the section of the throat immediately behind the mouth and above the larynx, trachea and oesophagus) gets longer. Consequently, the adult larynx doesn't go up as high relative to the mouth and thus is left exposed to food or drink falling on it. As noted earlier, if this kind of stuff enters our trachea, people can choke to death. It is necessary, therefore, to coordinate carefully the tongue, the larynx, a small flap called the epiglottis and the oesophageal sphincter (the round muscle in our food pipe) to avoid choking while eating. One thing people take care to avoid is talking with their mouths full. Talking and eating can kill or cause severe discomfort. Humans seem to have lost an advantage possessed by chimps and newborn humans.

But the news is not all bad. Although the full inventory of changes to the human vocal apparatus is too large and technical to discuss here, the final result of these developments enables us to talk more clearly than *Homo erectus*. This is because we can make a larger array of speech sounds, especially vowels, like the supervowels 'i', 'a' and 'u', which are found in all languages of the world. These are the easiest vowels to perceive. We're the only species that can make them well. Moreover, the vowel 'i' is of special interest. It enables the hearer to judge the length of the speaker's vocal tract and thus determine the relative size as well as the gender of the speaker and to 'normalise' the expectations for recognising that speaker's voice.

This evolutionary development of the vocal apparatus gives more options in the production of speech sounds, a production that begins with the lungs. The human lungs are to the vocal apparatus as a bottle of helium is to a carnival balloon. The mouth is like the airpiece. As air is released, the pitch of the escaping air sound can be manipulated by relaxing the piece, widening or narrowing the hole by which the air hisses out, cutting the air off intermittently and even 'jiggling' the balloon as the air is expelled.

But if human mouths and noses are like the balloon's airpiece, they also have more moving parts and more twists and chambers for the air to pass through than a balloon. So people can make many more sounds than a balloon. And since human ears and their inner workings have co-evolved with humans' sound-making system, it isn't surprising that they have evolved to make and be sensitive to a relatively narrow set of sounds that are used in speech.

According to evolutionary research, the larynges of all land animals evolved from the same source – the lung valves of ancient fish, in particular as seen in the Protopterus, the Neoceratodus and the Lepidosiren. Fish gave speech as we know it. The two slits in this archaic fish valve functioned to prevent water entering into the lungs of the fish. To this simple muscular mechanism, evolution added cartilage and tinkered a bit more to allow for mammalian breathing and the process of phonation. Our resultant vocal cords are therefore actually a complex set of muscles. They were first called *cordes* by the French researcher, Ferrein, who conceived of the vocal apparatus as a musical instrument.[8]

What is complicated, on the other hand, is the control of this device.

Humans do not play their vocal apparatuses by hand. They control each move of hundreds of muscles, from the diaphragm to the tongue to the opening of the naso-pharyngeal passage, with their brains. Just as the shape of the vocal apparatus has changed over the millennia to produce more discernible speech, matching more effectively the nuances of the language that speakers have in their heads, so the brain evolved connections to control the vocal apparatus.

Humans must be able to control their breathing effectively to produce speech. Whereas breathing involves inspiration and expiration, speech is almost exclusively expiration. This requires control of the flow of air and the regulation of air-pressure from the lungs through the vocal folds. Speech requires the ability to keep producing speech sounds even after the point of 'quiet breathing' (wherein air is not forced out of the lungs in exhalation by normal muscle action, but allowed to seep out of the lungs passively). This control enables people to speak in long sentences, with the attendant production not only of individual speech sounds, such as vowels and consonants, but also of the pitch and modulation of loudness and duration of segments and phrases.

It is obvious that the brain has a tight link to vocal production because electrical stimulation of parts of the brain can produce articulatory movements and some examples of phonation (vowel sounds in particular). Other primate brains respond differently. Stimulation of the regions corresponding to Brodmann Area 44 in other primates produces face, tongue and vocal cord movements but not phonation as similar stimulation produces in humans.

To state the obvious, chimpanzees are unable to talk. But this is not, as some claim, because of their vocal tract. A chimp's vocal tract certainly could produce enough distinct sounds to support a language of some sort. Chimps do not talk, rather, because of their brains – they are not intelligent enough to support the kind of grammars that humans use and they are not able to control their vocal tracts finely enough to control speech production. Lieberman locates the main controllers of speech in the basal ganglia, what he and others refer to as our reptilian brain. The basal ganglia are, again, responsible for habit-like behaviours among others. Disruption of the neural circuits linking the basal ganglia to the cortex can result in disorders such as obsessive-compulsive disorder, schizophrenia and Parkinson's disease. The basal ganglia are implicated in motor control, aspects of cognition, attention and several other aspects of human behaviour.

Therefore, in conjunction with the evolved form of the FOXP2 gene, which allows for better control of the vocal apparatus and mental processing of the kind used in modern humans' language, the evolution of connections between the basal ganglia and the larger human cerebral cortex are essential to support human speech (or sign language). Recognising these changes helps us to recognise that human language and speech are part of a continuum seen in several other species. It is not that there is any special gene for language or an unbridgeable gap that appeared suddenly to provide humans with language and speech. Rather, what the evolutionary record shows is that the language gap was formed over millions of years by baby steps. At the same time, *erectus* is a fine example of how early the language threshold was crossed, how changes in the brain and human intelligence were able to offer up human language even with ape-like speech capabilities. *Homo erectus* is the evidence that apes could talk if they had brains large enough. Humans are those apes.

Part Three

The Evolution of Language Form

9

Where Grammar Came From

Speech is a non-instinctive, acquired, 'cultural' function.

Edward Sapir

SOMEONE MIGHT ASK this in English: 'Yesterday, what did John give to Mary in the library?' And someone else might answer, '*Catcher in the Rye.*'

This is a complete conversation. Not a particularly fecund one, but, still, typical of the exchanges that people depend on in their day-to-day lives, representative of the way that brains are supersized by culture as well as the role of language in expanding knowledge from the brain of a single individual to the shared knowledge of all the individuals of a society, of all the individuals alive, in fact. Even of all the individuals who have ever lived and written or been written about. It wasn't the computer that ushered in the information age, it was language. The information age began nearly 2 million years ago. *Homo sapiens* have just tweaked it a bit.

Discourse and conversation are the apex of language. In what way, though, is this apical position revealed in the sentences in the discourse above? When native speakers of English hear the first sentence of our conversation in a natural context, they understand it. They are able to do this because they have learned to listen to all the parts of this complex whole and to use each part to help them understand what the speaker intends when they ask, 'Yesterday, what did John give to Mary in the library?'

First, they understand the words, 'library', 'in', 'did', 'John,' and the rest. Second, all English speakers will hear the word with the greatest amplitude or loudness and also notice which words receive the highest

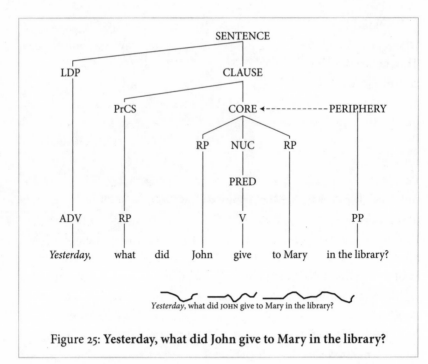

Figure 25: **Yesterday, what did John give to Mary in the library?**

and lowest pitches. The loudness and pitch can vary according to what the speaker is trying to communicate. They aren't always the same, not even for the same sentence. Figure 25 shows one way to assign pitch and loudness to our sentence.*

The line above the words shows the melody, the relative pitches, over the entire sentence. The italicised 'yesterday' indicates that it is the second-loudest word in the sentence. The small caps of 'John' mean that that is the loudest word in the sentence. 'Yesterday' and 'John' are selected by the speaker to indicate that they warrant the hearer's particular attention. The melody indicates that this is a question, but it also picks out the four words 'yesterday', 'John', 'Mary' and 'library' for higher pitches, indicating that they represent different kinds of information needed to process the request.

*This diagram comes from the linguistic theory known as *Role and Reference Grammar,* one of the only theories of language in existence that concerns itself with the whole utterance, not just its syntax.

The loudness and pitch of 'yesterday' says this – one is talking about what someone gave to someone yesterday, not today, not another day. This helps the hearer avoid confusion. The word 'yesterday' doesn't communicate this information by itself. The word is aided by pitch and loudness, which highlight its specialness for both the information being communicated and the information being requested. 'John' is the loudest word because it is particularly important to the speaker for the hearer to tell him what 'John' did. Maybe Mary is a librarian. People give her books all day, every day. Then what is being asked is not about what Suzy gave her but only what John gave her. The pitch and loudness let the hearer know this so that they do not have to sort through all the people who gave Mary things yesterday. The word 'John' already makes this clear. The pitch and loudness highlight this. They offer additional cues to the hearer to guide their mental search for the right information.

Now, when someone asks this question, what does the asker look like? Probably they look something like this: the upper arms are held against their sides, with their forearms extended, palms facing outward and their brow furrowed. These body and hand expressions are important. The hearer uses these gestural, facial and other body cues to know immediately that you are not making a statement. A question is being asked.

Now consider the words themselves. The word 'yesterday' appears at the far left of the sentence. In the sentence below the <>'s indicate the other places that 'yesterday' could appear in the sentence:

<> what <> did John <> give <> to Mary <> in the library <>?

So we could ask:

'What did John give yesterday to Mary in the library?'
'Yesterday what did John give to Mary in the library?'
'What did John, yesterday, give to Mary in the library?'
'What did John give to Mary yesterday in the library?'
'What did John give to Mary in the library yesterday?'
'What, yesterday, did John give to Mary in the library?'

But a native speaker is less likely to ask the question like this:

*'What did yesterday John give to Mary in the library?'
*'What did John give to yesterday Mary in the library?'
*'What did John give to Mary in yesterday the library?'
*'What did John give to Mary in the yesterday library?'

The asterisk in the examples just above means that one is perhaps less likely to hear these sentences other than in a very specific context, if at all. This exercise could be continued with other words or phrases of the sample sentence:

'In the library, *what did John give to Mary yesterday*?'

This exercise on words and their orders need not be continued, however, since there is now enough information to know that putting together a sentence is not simply stringing words like beads on a cord.

Part of the organisation of the sentence is the grouping of words in most languages into phrases. These phrases should not be interrupted, which is why one cannot place yesterday after the word 'in' or after the word 'the'. Grammatical phrases are forms of 'chunking' for short-term memory. They aid recall and interpretation.

Still, a great deal of cultural information is missing from the above example. For instance what in the hell is a library? Is John a man or a woman? Is Mary a man or a woman? Which library is being talked about? What kinds of things are most likely to be given by John? Do John and Mary know each other? Although there are a lot of questions, one quickly perceives the answers to them if they are the speaker or hearer in the above conversation because of the knowledge people absorb – often without instruction – from their surrounding society and culture. People use individual knowledge (such as which library is the most likely candidate) and cultural knowledge (such as what a library is) to narrow down their 'solution space' as hearers. They are not, therefore, obligated to sort through all possible information to understand and respond, but must only mentally scroll through the culturally and individually most pertinent information that might fit the question at hand. The syntax, word selection, intonation and amplitude are all designed to help understand what was just said.

But there's more, such as certain kinds of information in the example here. There is shared information, signalled sometimes by words such as 'the' in the phrase 'the library'. Because someone says '*the* library' and not '*a* library', they signal to the hearer that this is a library that both know – they share this knowledge – because of the context in which they are having this conversation. The request here is for new information, as signalled by the question-word 'what'. This is information that the speaker does not have but expects the hearer to have. Sentences exist to facilitate information exchange between speakers. Grammar is simply a tool to facilitate that.

The question is also an intentional act – an action intended to elicit a particular kind of action from the hearer. The action desired here is 'give me the information I wish or tell me where to get it'. Actions vary. Thus, if a king were to say, 'Off with his head,' the action desired would be a decapitation, if he were being literal. Literalness brings us to another twist on how sentences are uttered and understood – is the speaker being literal or ironic or figurative? Are they insane?

With language, speakers can recognise promises, declarations, indirect requests, direct requests, denunciations, legal impact ('I now pronounce you man and wife') and other culturally significant information points. No theory of language should neglect to tell us about the complexity of language and how its parts fit together – intonation, gestures, grammar, lexical choice, type of intention and the rest. And what does the hearer do in this river of signals and information? Does she sit and ponder for hours before giving an answer? No she understands all of this implicitly and instantly. These cues work together. Taken as one, they make the sentence easier to understand, not harder. And the evidence is that the single strongest force driving this instantaneous comprehension is information structure. What is new? What is shared? And this derives not merely from the literal meanings of the words but from that implicit cultural knowledge that I call dark matter.

Into the syntactic representation of a sentence, speakers intercalate gestures and intonation. They use these gestures as annotations to indicate the presence of implicit information from the culture and the personal experiences of the speaker and hearer. But something is always left out. The language never expresses everything. The culture fills in the details.

How did human languages go from simple symbols to this complex interaction between higher-level symbols, symbols within symbols, grammar, intonation, gesture and culture? And why does all of this vary so much from language to language and culture to culture? Using the same words in British English or Australian English or Indian English or American English will produce related but distinct presupposed knowledge, intonational patterns, gestures and facial expressions. Far from there being any 'universal grammar' of integrating the various aspects of a single utterance, each culture largely follows its own head.

There are universally shared aspects of language, of course. Every culture uses pitch, attaches gestures and orders words in some agreed-upon sequence. These are necessary limits and features of language because they reflect physical and mental limitations of the species. Maybe – and this is exciting to think about – some of this represents vestiges of the ways that *Homo erectus* talked. Maybe humans passed a lot of grammar down by example, from millennium to millennium as the species continued to evolve. It is possible that modern languages have maintained 2-million-year-old solutions to information transfer first invented by *Homo erectus*. This possibility cannot be dismissed.

Reviewing what we have learned about symbols, these are based on a simple principle – namely that an arbitrary form can represent a meaning. Each symbol also entails Peirce's interpretant. Signs in all their forms are a first step towards another essential component of human language, the triple patterning of form and meaning via the addition of the interpretative aids of gestures and intonation. As symbols and the rest become more enculturated, they advance up the ladder from communication to language, to a distinction between the perspective of the outsider vs the perspective of the insider, what linguist Kenneth Pike referred to as the 'etic' (outsider viewpoint) and the 'emic' (insider viewpoint). Signs alone do not get us all the way to the etic and the emic. Culture is crucial all along the way.

The etic perspective is the perspective a tourist might have listening to a foreign language for the first time. 'They talk too fast.' 'I don't know how they understand one another with all those weird sounds.' But as one learns to speak the language, the sounds become more familiar, the language doesn't sound as though it is spoken all that rapidly after all, the language and its rules and pronunciation patterns become familiar.

The learner has travelled from the etic perspective of the outsider to the emic perspective of the insider.

By associating meanings with forms to create symbols, the distinction between form and meaning is highlighted.* And because symbols are interpreted by members of a particular group, they lead to the insider interpretation vs the outsider interpretation. This is what makes languages understandable to native speakers, but difficult to learn for non-native speakers. The progression to language is just this: Indexes → Icons → (emic) Symbols + (emic) grammar, (emic) gestures and (emic) intonation.

Following symbols, another important invention for language is grammar. Structure is needed to make more complex utterances out of symbols. A set of organising principles is required. These enable us to form utterances efficiently and most in line with the cultural expectations of the hearers.

Grammars are organised in two ways at once – vertically, also known as paradigmatic organisation, and horizontally, referred to as syntagmatic organisation. These modes of organisation underlie all grammars, as was pointed out at the beginning of the twentieth century by Swiss linguist Ferdinand de Saussure. Both vertical and horizontal organisation of a grammar work together to facilitate communication by allowing more information to be packed into individual words and phrases of language than would be possible without them. These modes of organisation follow from the nature of symbols and the transmission of information.

If one has symbols and sounds then there is no huge mental leap required to put these in some linear order. Linguists call the placing together of meaningless sounds ('phonemes' is the name given to speech sounds) into meaningful words 'duality of patterning'. For example the s, a and t of the word 'sat' are meaningless on their own. But assembled in this order, the word they form does have meaning. To form words, phonemes are taken from the list of a given language's sounds and placed into 'slots' to form a word, as, again, in 'sat': $s_{slot\ 1}\ a_{slot\ 2}\ t_{slot\ 3}$.

And once this duality becomes conventionalised, agreed upon by

---

*Gestures are also crucial to understanding how duality and compositionality happen.

Vertical/Paradigmatic				
	Symbol$_{slot1}$	Symbol$_{slot2}$	Symbol$_{slot3}$	
Symbol$_{filler 1}$ (John)	*John*	*saw*	*Mary*	
Symbol$_{filler 2}$ (Mary)				
Symbol$_{filler 3}$ (saw)				

Horizontal/Syntagmatic

Figure 26: **Extended duality of patterning – making a sentence**

the members of a culture, then it can be extended to combine meaningful items with meaningful items. From there, it is not a huge leap to use symbols for events and symbols for things together to make statements. Assume that one has a list of symbols. This is one aspect of the vertical or paradigmatic aspect of the grammar. Next there is an order to place these symbols in that a culture has agreed upon for the organisation of the symbols. So the task in forming a sentence or a phrase is to choose a symbol and place it in a slot, as is illustrated in Figure 26.

Knowing the grammar, which every speaker must, is just knowing the instructions for assembling the words into sentences. The simple grammar for this made-up language might just be: select one paradigmatic filler and place it in appropriate syntagmatic slot.

From the idea that there is an inventory of symbols to be placed in a specific order, not a huge jump cognitively, early humans would have been using the ideas of 'slot' and 'filler'. These are the bases of all grammars.

All of this was first explained by linguist Charles Hockett in 1960.[1] He called the combination of meaningless elements to make meaningful ones 'duality of patterning'. And once a people have symbols plus duality of patterning, then they extend duality to get the syntagmatic and paradigmatic organisation in the chart above. This almost gets us to human language. Only two other things are necessary – gestures and intonation. These together give the full language – symbols plus gestures and intonation. Here, though, the focus is on duality of patterning. As people organise their symbols they naturally begin next to analyse their symbols into smaller units. Thus a word such as 'cat', a symbol, is organised horizontally, or syntagmatically, as a syllable, c-a-t. But with this organisation it also becomes clear that 'cat' is organised vertically at the same time. So one could substitute a 'p' for the 'c' of cat to produce the word pat. Or one could substitute 'd' for 't' and get instead cad. In other words, 'cat' has three slots, c-a-t, and fillers for each slot come from the speech sounds of English.

The syllable is therefore itself an important part of the development of duality of patterning. It is a natural organising constraint set on the arrangement of phonemes that works to enable each phoneme to be better perceived. It has other functions, but the crucial point is that it is primarily an aid to perception, arising from the matching of ears to vocal apparatus over the course of human evolution, rather than a prespecified mental category. A very simple characterisation of the syllable is that speech sounds are arranged in order. The order preferred most of the time is that, from left to right in the syllable, the sounds are arranged from least inherently loud to the loudest and then back to softest. This makes the sounds in each syllable easier to hear. It is another way of chunking that helps our brains to keep track of what is going on in language. This property is called *sonority*. In simple terms, a sound is more sonorous if it is louder. Consonants are less sonorous than vowels. And among consonants some (these need not worry us here) are less sonorous than others.* Thus syllables are units of speech in which the individual slots produce a crescendo-decrescendo effect, where the nucleus or central part is the most sonorous element – usually a vowel – while at the margins are the least sonorous elements. This is shown by the syllable *bad*. This is an acceptable syllable in English because *b* and *d* are less sonorous than *a* and are found in the margins of the syllable while *a* – the most sonorous element – is in the nuclear or central position. The syllable *bda*, on the other hand, would be ill-formed in English because a less sonorous sound, *b*, is

---

*In *Dark Matter of the Mind* I offer a sustained discussion of phonology related to Universal Grammar, and I severely criticise the notion that either sonority or phonology is an innate property of human minds.

A commonly used representation of the 'sonority hierarchy' is: [a] > [e o] > [i u] > [r] > [l] > [m n ŋ] > [z v ð] > [s f θ] > [b d g] > [p t k].

followed by another less sonorous consonant sound, *d*, rather than immediately by an increase in sonority. This makes it hard to hear, or hard to distinguish *b* and *d* when they are placed together in the margin of the syllable.

Languages vary tremendously in syllabic organisation.* Certain ones, like English, have very complicated syllable patterns. The word 'strength' has more than one consonant in each margin. And the consonant 's' should follow the consonant 't' at the beginning of 'strength' because it is more sonorant. So the word should actually be 'tsrength'. It does not take this form because English has historically preferred the order 'st', based on sound patterns of earlier stages of English and the languages that influenced it, as well as cultural choice. History and culture are common factors that override and violate the otherwise purely phonetic organisation of syllables.

This perceptual and articulatory organisation of the syllable brings duality of patterning to language naturally. By organising sounds so that they can be more easily heard, languages in effect get this patterning for free. Each margin and nucleus of a syllable is part of the horizontal organisation of the syllable while the sounds that can go into the margin or nucleus are the fillers. And that means that the syllable in a sense could have been the key to grammar and more complex languages. Again, the syllable is based upon a simple idea: 'chunk sounds so as to make them easier to hear and to remember'. *Homo erectus* would have probably have had syllables because they just follow from the shortness of our short-term memory in conjunction with the easiest arrangements to hear. If so, then this means that *erectus* would have had grammar practically given to them on a platter as soon as they used syllables. It is, of course, possible that syllables came later in language evolution than, say, words and sentences, but any type of sound organisation, whether in phonemes of *Homo sapiens* or other sounds by *erectus* or *neanderthalensis*, would entail a more powerful form for organising a language and moving it beyond mere symbols to some form of grammar. Thus early speech would have been a stimulus to syntagmatic and paradigmatic organisation in syntax, morphology

---

*I have written extensively on syllables in Amazonian languages, which are particularly interesting theoretically.

and elsewhere in language. In fact, some other animals, such as cotton-top tamarins, are also claimed to have syllables. Whatever tamarins can do, the bet is that *erectus* could do better. If tamarins had better-equipped brains then they'd be on their way to human language.

If this is on the right track, duality of patterning, along with gestures and intonation, are the foundational organisational principles of language. However, once these elements are in place, it would be expected that languages would discover the utility of hierarchy, which computer scientists and psychologists have argued to be ever useful in the transmission or storage of complex information.

Phonology is, like all other forms of human behaviour, constrained by the memory–expression tension: the more units that a language has, the less ambiguously it can express messages, but the more there is to learn and memorise. So if a language had three hundred speech sounds, it could produce words that were less ambiguous than a language with only five speech sounds. But this has a cost – the language with more speech sounds is harder to learn. Phonology organises sounds to make them easier to perceive, adding a few local cultural modifications preferred by a particular community (as English 'strength' instead of 'tsrength'). This gets to the co-evolution of the articulatory and auditory apparatuses. The relationship between humans' ears and their mouths is what accounts for the sounds of all human languages. It is what makes human speech sounds different than, say, Martian speech sounds.

The articulatory apparatus of humans is, of course, also interesting because no single part of it – other than its shape – is specialised for speech. As we have seen, the human vocal apparatus has three basic components – moving parts (the articulators), stationary parts (the points of articulation), and airflow generators. It is worth underscoring again the fact that the evolution of the vocal apparatus for speech is likely to have followed the beginning of language. Though language can exist without well-developed speech abilities (many modern languages can be whistled, hummed, or signed), there can be no speech without language. *Neanderthalensis* did not have speech capabilities like those of *sapiens*. But they most certainly could have had a working language without a sapiens vocal apparatus. The inability of *neanderthalensis* to produce /i/, /a/ and /u/ (at least according to Philip Lieberman) would be a handicap for speech, but these 'cardinal' or 'quantal' vowels are neither necessary nor sufficient for language (not necessary because of signed languages, not sufficient because parrots can produce them).

Speech is enhanced as has been mentioned, when the auditory system co-evolves with the articulatory system. This just means that human ears and mouths evolve together. Humans therefore get better at hearing the sounds their mouths most easily make and making the sounds their ears most easily perceive.

Individual speech sounds, phones, are produced by the articulators – tongue and lips for the most part – meeting or approximating points of articulation – alveolar ridge, teeth, palate, lips and so on. Some of these sounds are louder because they offer minimal impedance to the flow of air out of the mouth (and for many out of the nose). These are vowels. No articulator makes direct contact with a point of articulation in the production of vowels. Other phones completely or partially impede the flow of air out of the mouth. These are consonants. With both consonants and vowels the stream of sounds produced by any speaker can be organised so as to maximise both information rate (consonants generally carry more information than vowels, since there are more of them) and perceptual clarity (consonants are easier to perceive in different positions of the speech stream, such as immediately preceding and following vowels and the beginnings and ends of words). Vowels and consonants, since speech is not digital in its production, but rather a continuous stream of articulatory movements, 'assimilate' to one another, that is they become more alike in some contexts, though not always the same contexts in every language. If a native speaker of English utters the word 'clock', the 'k' at the end is pronounced further back in the mouth than it is when they pronounce the word 'click'. This is because the vowel 'o' is further back in the mouth and the vowel 'i' is further to the front of the mouth. In these cases, the vowel 'pulls' the consonant towards its place of articulation. Additional modifications to sounds enhance perception of speech sounds. Another example is known as aspiration – a puff of air made when producing a sound. Or voicing – what happens as the vocal cords vibrate while producing a sound. Syllable structure is another modification, when sounds are pronounced differently in different positions of a syllable. This is seen in the pronunciation of 'l' when it is at the end of a syllable, as in the word 'bull', vs 'l' when it is at the beginning of a syllable as in 'leaf'. To literally see aspiration, hold a piece of paper about one inch in front of your mouth and pronounce the word 'paper'. The paper will move. Now do the same while pronouncing the word 'spa'. The paper will not move on the 'p' of 'spa' if one is a native speaker of English.

These enhancements are ignored often by native speakers when they produce speech because such enhancements are simply 'add-ons' and not part of the target sound. This is why native speakers of English do not normally hear a difference between the [p] of 'spa' and the [pʰ] of 'paper', where the raised 'ʰ' following a consonant indicates aspiration. But to the linguist, these sounds are quite distinct. Speakers are unaware of such enhancements and usually can learn to hear them only with special effort. The study of the physical properties of sounds, regardless of the speakers' perceptions and organisation of them, is phonetics. The study of the emic knowledge of speakers, what enhancements are ignored by native speakers and what sounds they target, is phonology.

Continuing on with our study of phonology, there is a long tradition that breaks basic sounds, vowels and consonants, down further into phonetic features, [+/– voiced] 'voiced vs non-voiced' or [+/– advanced tongue root], as in the contrast between the English vowels /i/ of beet and the vowel i of [bit] and so on. But no harm is done to the exposition of language evolution if such finer details are ignored.

Moving up the phonological hierarchy we arrive once again at the syllable 'the', which introduces duality of patterning into speech sound organisation. To elaborate slightly further on what was said earlier about the syllable, consider the syllables in Figure 27.

As per the earlier discussion of sonority, it is expected that the syllable [sat] will be well formed, ceteris paribus, while the syllable [lbad] will not be because in the latter the sounds are harder to perceive.

The syllable is thus a hierarchical, non-recursive structuring of speech sounds. It functions to enhance the perceptibility of phones and often works in languages as the basic rhythmic unit. Once again, one can imagine that, given their extremely useful contributions to speech perception, syllables began to appear early on in the linking of sounds to meaning in language. They would have been a very useful and easy

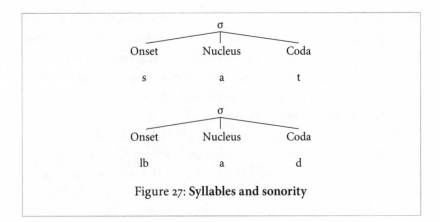

Figure 27: **Syllables and sonority**

addition to speech, dramatically improving the perceptibility of speech sounds. The natural limitations of human auditory and articulatory systems would have exerted pressure for speakers to hear and produce syllabic organisation early on.

However, once introduced, syllables, segments and other units of the phonological hierarchy would have undergone culture-based elaborations. Changes, in other words, are made to satisfy local preferences, without regard for ease of pronunciation or production. These elaborations are useful for group identification, as well as the perceptions of sounds in certain places in the words. So sometimes they are motivated by ease of hearing or pronunciation or by cultural reasons, in order to make sounds that identified one group as the source of those sounds, because speakers of one culture would have preferred some sounds to others, some enhancements to others and so on. A particular language's inventory of sounds might also be culturally limited. This all means that in the history of languages, a set of cultural preferences emerges that selects among the sounds that humans can produce and perceive to choose the sounds that a particular culture at a particular time in the development of the language chooses to use. After this selection, the preferred sounds and patterns will change over time, subject to new articulatory, auditory, or cultural pressures, or via contact with other languages.

Other units of the phonological hierarchy include phonological phrases, groupings of syllables into phonological words or units larger than words. These phrases or words are also forms of chunking to aid working memory and to facilitate faster interpretation of information communicated. This segmentation is aided by gestures and intonation which offer backup help to speech for perception and working memory. In this way, the grouping of smaller linguistic units (such as sounds) into larger linguistic units (such as syllables and words and phrases) facilitates communication. Phrases and words are themselves grouped into larger groupings some linguists have referred to as 'contours' or 'breath groups' – groupings of sounds marked by breathing and intonation. We have mentioned that pitch, loudness and the lengthening or shortening of some words or phrases can be used to distinguish, say, new information from old information, such as the topic we are currently discussing (old information) and a comment about that topic (new information). These can also be used to signal emphasis and other nuances that the speaker would like the hearer to pick up on about what is being communicated. All of these uses of phonology emerge gradually as humans move from indexes to grammar. And every step of the way they would have in all likelihood been accompanied by gesture.

From these baby steps an entire 'phonological hierarchy' is constructed. This hierarchy entails that most elements of the sound structure of a given language are composed of smaller elements. In other words, each unit of sound is built up from smaller units via natural processes that make it easier to hear and produce the sounds of a given language. The standard linguistic view of the phonological hierarchy is given in Figure 28.

Our sound structures are also constrained by two other sets of factors. The first is the environment. Sound structures can be significantly constrained by the environmental conditions in which the language arose – average temperatures, humidity, atmospheric pressure and so on. Linguists missed these connections for most of the history of language, though more recent research has now established them clearly. Thus, to understand the evolution of a specific language, one must know something about both its original culture and its ecological circumstances. No language is an island.

Several researchers summarise these generalisations with the proposal that the first utterances of humans were 'holophrastic'. That is,

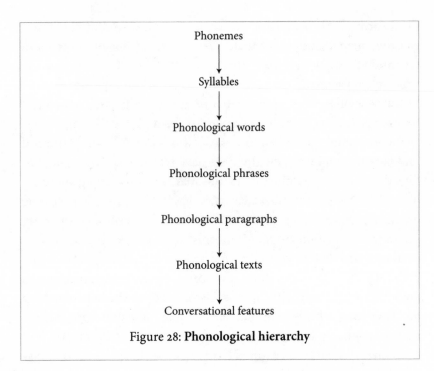

Figure 28: **Phonological hierarchy**

the first attempts at communication were unstructured utterances that were neither words nor sentences, simply interjections or exclamations. If an *erectus* used the same expression over and over to refer to a sabre-toothed cat, say, how might or would that symbol be decomposed into smaller pieces? Gestures, with functions that overlap intonation in some ways, contribute to decomposing a larger unit into smaller constituents, either reinforcing already highlighted portions or by indicating that other portions of the utterance are of secondary importance, but still more than portions of tertiary importance, and so on. To see how this might work, imagine that an *erectus* woman saw a large cat run by her and she exclaimed, 'Shamalamadingdong!' One of those syllables or portions of that utterance might be louder or higher pitched than the others. If she were emotional, that would necessarily come through in gestures and pitches that would intentionally or inadvertently highlight different portions of that utterance. Perhaps such as 'SHAMAlama-dingDONG!' or 'ShamaLAMAdingdong' or 'ShamalamaDINGdong' or 'SHAMAlamaDINGdong', and so forth. If her gestures, loudness, pitch (high or low or medium) lined up with the same syllables, then those would perhaps begin to get recognition as parts of a word or sentence that began without any parts.

Prosody (pitch, loudness, length), gestures and other markers of salience (body positioning, eyebrow raising, etc.) have the joint effect of beginning to decompose the utterance, breaking it down into parts according to their pitch or gestures. Once utterances are decomposed, and only then, they can be (re)composed (synthesised), to build additional utterances. And this leads to another necessary property of human language: semantic compositionality. This is crucial for all languages. It is the ability to encode or decode a meaning of a whole utterance from the individual meanings of its parts.

Thus from natural processes linking sound and meaning in utterances, there is an easy path via gestures, intonation, duration and amplitude to decomposing an initially unstructured whole into parts and from there recomposing parts into wholes. And this is the birth of all grammars. No special genes required.

Kenneth Pike placed morphology and syntax together in another hierarchy, worth repeating here (though I use my own, slightly adapted version). This he called the 'morphosyntactic hierarchy' – the building up of conversations from smaller and smaller parts.

How innovations, linguistic or otherwise, spread and become part of a language is a puzzle known as the 'actuation problem'. Just as with the spread of new words or expressions or jokes today, several possible enabling factors might be involved in the origin and spread of linguistic innovations. Speakers might like the sounds of certain components of novel *erectus* utterances more than others, or an accompanying pitch and/or gesture might have highlighted one portion of the utterance to the exclusion of other parts. As highlighting is also picked up by others and begins to circulate, for whatever reasons, the highlighted, more salient portion becomes more important in the transmission and perception of the utterance being 'actuated'.

It is likely that the first utterance was made to communicate to someone else. Of course, there are no witnesses. Nevertheless the prior and subsequent history of the language progression strongly support this. Language is about communication. The possibly clearer thinking that comes when we can think in speech instead of merely, say, in

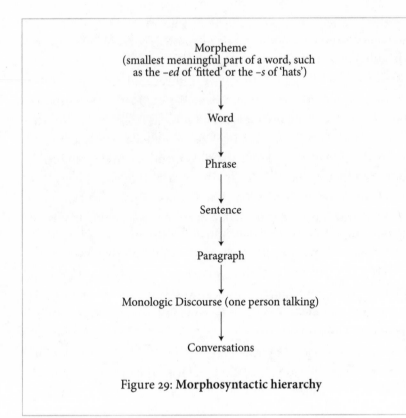

Figure 29: **Morphosyntactic hierarchy**

pictures, is a by-product of language. It is not itself what language is about.

Just as there is no need to appeal to genes for syntax, the evidence suggests that neither are sound structures innate, aside from the innate vocal apparatus–auditory perception link between what sounds people can best produce and hear. The simplest hypothesis is that the co-evolution of the vocal apparatus, the hearing apparatus and linguistic organisational principles led to the existence of a well-organised sound-based system of forms for representing meanings as part of signs. There are external, functional and ecological constraints on the evolution of sound systems.*

---

*Sign languages do not have phonologies, except in a metaphorical sense, though they do organise gestures in ways reminiscent of sound structures. Fully developed

Syntax develops with duality of patterning and additions to it that are based on cultural communicational objectives and conventions, along with different grammatical strategies. This means that one can add recursion if it is a culturally beneficial strategy. One can have relative clauses or not. One can have conjoined noun phrases or not. Some examples of different grammar strategies in English include sentences like:

John and Mary went to town (a complex, conjoined noun phrase) vs John went to town. Mary went to town (two simple sentences).

The man is tall. The man is here (two simple sentences). vs The man *who is tall* is here (a complex sentence with a relative clause).

Morphology is the scientific term for how words are constructed. Different languages use different word-building strategies, though the set of workable strategies is small. Thus in English there are at most five different forms for any verb: *sing, sang, sung, singing, sings*. Some verbs have even fewer forms: *hit, hitting, hits*. There are really only a few choices for building morphological (word) structures. Words can be simple or composed of parts (called morphemes). If they are simple, without internal divisions, this is an 'isolating' language. Chinese is one illustration. In Chinese there is usually no past tense form for a verb. A separate word is necessary to indicate past tense (or just the context). So where we would say in English, 'I ran' (past tense) vs 'I run' (present tense), in Chinese you might say, 'I now run' (three words) vs 'I yesterday run' (three words).

In another language, such as Portuguese, the strategy for building words is different. Like many Romance languages (those descended from Latin), Portuguese words can combine several meanings. A simple example can be constructed from **falo**, which means 'I speak' in Portuguese.

The 'o' ending of the verb means several things at once. It indicates first person singular, 'I'. But it also means 'present tense'. And

---

sign languages usually arise when phonologies are unavailable (through deafness or lack of articulatory ability) or when other cultural values render gestures preferable. Since gestures are related to the eyes rather than the ears, their organising principles are different in some ways. Of course, because both phonological and gestural languages are designed by cultures and the minds, subject to similar constraints of computational utility, they will also have features in common, as is often observed in the literature.

it also means 'indicative mood' (which in turn means, very roughly, 'this really is happening'). The 'o' of **falo** also means that the verb is of the -**ar** set of verbs (**falar** 'to speak', **quebrar** 'to break', **olhar** 'to look', and many, many others). Portuguese and other languages descended from Latin, such as Spanish, Romanian, Italian, are known as Romance languages. They are referred to linguistically as 'inflectional' languages.

Other languages, such as Turkish and many Native American languages, are called 'agglutinative'. This means that each part of each word often has a single meaning, unlike Romance languages, where each part of the word can have several meanings, as in the 'o' of **falo**. A single word in Turkish can be long and have many parts, but each part has but a single meaning:

Çekoslovakyalılaştıramadıklarımızdanmışsınızcasına 'as if you were one of those whom we could not make resemble the Czechoslovakian people'.

Some inflectional languages even have special kinds of morphemes called circumfixes. In German the past tense of the verb **spielen** 'to play' is **gespielt**, where **ge-** and **-t** jointly express the past tense, circumscribing the verb they affect.

Other languages use pitch to add meaning to words. This produces what are called simulfixes. In Pirahã the nearly identical words **ʔáagá** 'permanent quality' vs **ʔaagá** 'temporary quality' are distinguished only by the high tone on the first vowel **á**. So I could say, **Ti báaʔáí ʔáagá** 'I am nice (always)' or, **Ti báaʔáí ʔaagá** 'I am nice (at present)'.

Or one could express some meaning on the consonants and another part of the meaning of the word on the vowels, in which case we have a non-concatenative system. Arabic languages are of this type. But English also has examples. So **foot** is singular, but **feet** is plural, where we have kept the consonants **f** and **t** the same but changed the vowels.

It is unlikely that any language has a 'pure' system of word structure, using only one of these strategies just exemplified. Languages tend to mix different approaches to building words. The mixing is often caused by historical accidents, remnants of earlier stages of the language's development or contact with other languages. But what this brief summary of word formation shows is that if we look at all the

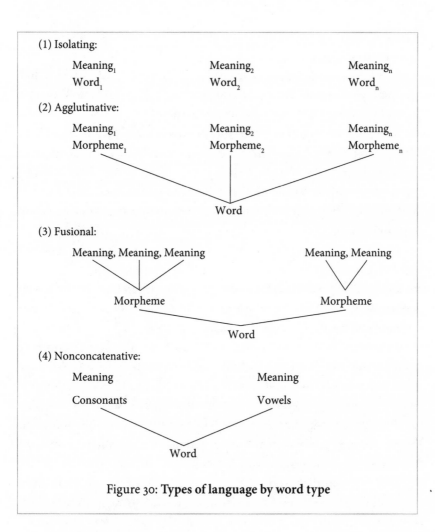

Figure 30: **Types of language by word type**

morphological systems of the world the basic principles are easy. This is summarised in Figure 30.*

These are the choices that every culture has to make. It could choose multiples of these, but simplicity (for memory's sake) will favour less

---

*One example of a a nonconcatenative process comes from English, where 'foot' is singular and 'feet' is plural, with one vowel substituting for another. Another famous set of cases comes from Semitic languages, such as Hebrew, as seen in the example below:

'he dictated' (causative active v.) is TB?n 'hih'tiv', while 'it was dictated' (causative passive v.) is nrqri 'huh'tav' rather than more mixing of the systems. From the beginnings of the invention of symbols, at least 1.9 million years ago by *Homo erectus*, there has been sufficient time to discover this small range of possibilities and, via the grammar, meaning, pitches and gestures of language, to build morphological systems from them.*

Arguably the greatest contribution Noam Chomsky has made to the understanding of human languages is his classification of different kinds of grammars, based on their computational and mathematical properties.[2] This classification has become known as the 'Chomsky hierarchy of grammars', though it was heavily influenced by the work of Emile Post and Marcel Schützenberger.

Yet while Chomsky's work is insightful and has been used for decades by computer scientists, psychologists and linguists, it denies that language is a system for communication. Therefore, in spite of its influence, it is ignored here in order to discuss a less complicated but arguably more effective way of looking at grammar's place in the evolution of language as a communicative tool. The claim is that, contrary to some theories, there are various kinds of grammars available to the world's languages and cultures (linear grammars, hierarchical grammars and recursive hierarchical grammars). These systems are available to all languages and they are the only systems for organising grammar. There are only three organisational templates for human syntax. In principle, this is not excessively difficult.

A linear grammar would be an arrangement of words left to right in a culturally specified order. In other words, linear grammars are not merely the stringing of words without thought. A language might

---

*There is some overlap between the account presented here and an independently developed set of proposals by Erkki and Hendrik Luuk: 'The Evolution of Syntax: Signs, Concatenation and Embedding' (*Cognitive Systems Research* 27, 2014: 1–10), a very important article in the literature on the evolution of syntax. Like the account below, Luuk and Luuk argue that syntax develops initially from the concatenation of signs, moving then from mere concatenation to embedding grammars. My differences with their proposals, however, are many. For one thing, they seem to believe, which is common enough, that compositionality depends on syntactic structure, failing to recognise that semantic compositionality is facilitated by syntactic structure, but not dependent on it in all languages. Further, they fail to recognise the *cultural context* and why therefore modern languages do not need embedding. They do seem to embrace my view that language is a cultural tool.

stipulate that the basic order of its words is subject noun + predicate verb + direct object noun, producing sentences like 'John$_{\text{subject noun}}$ hit-Bill$_{\text{direct object noun}}$' Or, if we look at the Amazonian language Hixkaryana, the order would be object noun + predicate verb + subject noun. This would produce 'Bill$_{\text{object noun}}$ hit$_{\text{predicate verb}}$ John$_{\text{subject noun}}$' and this sentence would be translated into English from Hixkaryana, in spite of its order of words, as 'John hit Bill.' These word orders, like all constituents in human languages, are not mysterious grammatical processes dissociated from communication. On the contrary, the data suggest that every bit of grammar evolves to aid short-term memory and the understanding of utterances. Throughout all languages heretofore studied, grammatical strategies are used to keep track of which words are more closely related.

One common strategy of linking related words is to place words closer to words whose meaning they most affect. Another (usually in conjunction with the first) is to place words in phrases hierarchically, as in one possible grammatical structure for the phrase *John's very big book* shown in Figure 31.

In this phrase there are chunks or constituents within chunks. The constituent 'very big' is a chunk of the larger phrase 'very big book'.

One way to keep track of which words are most closely related is to mark them with case or agreement. Greek and Latin allow words that are related to be separated by other words in a sentence, so long as they are all marked with the same case (genitive, ablative, accusative, nominative and so on).

Consider the Greek sentences below, transcribed in English orthography, all of which mean 'Antigone adores Surrealism':

(a)	latrevi	ton iperealismo	i Antigoni
	Verb	Object	Subject
	adore$_{\text{verb}}$	the surrealism$_{\text{accusative}}$	the Antigoni$_{\text{nominative}}$
(b)	latrevi i Antigoni$_{\text{nominative}}$	ton iperealismo$_{\text{accusative}}$	
	Verb	Subject	Object
	adore$_{\text{verb}}$	the Antigoni$_{\text{nominative}}$	the surrealism$_{\text{accusative}}$
(c)	i Antigoni	latrevi	ton iperealismo
	Subject	Verb	Object
	the Antigoni$_{\text{nominative}}$	adore$_{\text{verb}}$	the surrealism$_{\text{accusative}}$

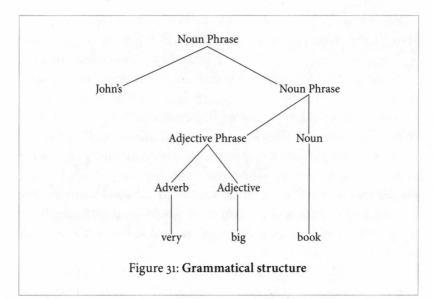

Figure 31: **Grammatical structure**

(d)　ton iperrealismo　　　latrevi　　　　　i Antigoni

　　　Object　　　　　　　　Verb　　　　　Subject

　　　the surrealism$_{accusative}$　adore$_{verb}$　the Antigoni$_{nominative}$

(e)　i Antigoni　　　　　　ton iperealismo　latrevi

　　　Subject　　　　　　　Object　　　　　Verb

　　　the Antigoni$_{nominative}$　the surrealism$_{accusative}$　adore$_{verb}$

(f)　ton iperealismo　　　i Antigoni　　　latrevi

　　　Object　　　　　　　Subject　　　　Verb

　　　the surrealism$_{accusative}$　the Antigoni$_{nominative}$　adore$_{verb}$

All of these sentences are grammatical in Greek. All are regularly found. The Greek language, like Latin and many other languages, allows freer word order than, say, English, because the cases, such as 'nominative' and 'accusative' distinguish the object from the subject. In English, since we rarely use case, word order, not case, has been adopted by our culture as the way to keep straight who did what to whom.

But even English has case to a small degree: I$_{nominative}$ saw him$_{objective}$, He$_{nominative}$ saw me$_{objective}$, but not *Me saw he or *Him saw I. English pronouns are marked for case ('I' and 'he' are nominative; 'me' and 'him' are objective).

We see agreement in all English sentences, as in 'He likes John.' Here, 'he' is third person singular and so the verb carries the third person singular 's' ending on 'likes' to agree with the subject, he. But if we say instead, 'I like John,' the verb is not 'likes' but 'like' because the pronoun 'I' is not third person. Agreement is simply another way to keep track of word relationships in a sentence.

Placing words together in linear order without additional structure is an option used by some languages to organise their grammars, avoiding tree structures and recursion. Even primates such as Koko the gorilla have been able to master such word order-based structures. It certainly would have been within the capabilities of *Homo erectus* to master such a language and would have been the first strategy to be used (and is still used in a few modern languages, such as perhaps Pirahã and the Indonesian language Riau).

One thing that all grammars must have is a way to assemble the meanings of the parts of an utterance to form the meaning of the whole utterance. How do the three words, 'he', 'John' and 'see', come to mean 'He sees John' as a sentence? The meaning of each of the words is put together in phrases and sentences. This is called compositionality, without which there can be no language. This property can simply be built off the phrases in a tree structure directly, which is the case for most languages, such as English, Spanish, Turkish and thousands of others. But it can also be built in a linear grammar. In fact, even English shows that compositionality doesn't need complex grammar. The utterance 'Eat. Drink. Man. Woman.' is perfectly intelligible. But as the meaning is put together, it relies heavily on our cultural knowledge. The absence of complex (or any) syntax doesn't scuttle the meaning.

Many languages, if not all, have examples of meaningful sentences with little or no syntax. (Another example from English would be 'You drink. You drive. You go to jail.' These three sentences have the same interpretation as the grammatically distinct and more complex construction, 'If you drink and drive, then you will go to jail.') Note that the interpretation of separate sentences as a single sentence does not require prior knowledge of other syntactic possibilities, since we are able to use this kind of cultural understanding to interpret entire stories in multiple ways. One might object that sentences without sufficient syntax are ambiguous – they have multiple meanings and confuse the hearer. But, as noted in chapter 4, even structured sentences like 'Flying planes can be dangerous' are ambiguous, in spite of English's elaborate syntax. One could eliminate or at least reduce ambiguity in a given language, but the means for doing so always include making the grammar more difficult or the list of words longer. And these strategies only make the language more complicated than it needs to be, by and large. So this is why, in all languages, ambiguous sentences are interpreted through native speaker knowledge of the context, the speaker and their culture.

The more closely the syntax matches the meaning, usually the easier the interpretation is. But much of grammar is a cultural choice. The form of a grammar is not genetically predestined. Linear grammars are not only the probable initial stage of syntax, but also still viable grammars for several modern languages, as we have seen. A linear grammar with symbols, intonation and gestures is all a G_1 language – the simplest full human language – requires.

The next type of languages are G_2 languages. These are languages that have hierarchical structures (like the 'tree diagrams' of sentences and phrases), but lack recursion. Some have claimed that Pirahã and Riau are examples. Yet we need not focus exclusively on isolated or rarely spoken languages such as these. One researcher, Fred Karlsson, claims that *most* European languages have hierarchy but not recursion.

Karlsson bases his claim on the observation: 'No genuine triple initial embeddings nor any quadruple centre-embeddings are on record ("genuine" here means sentences produced in natural non-linguistic contexts, not sentences produced by professional linguists in the course of their theoretical argumentation).'[3] 'Embedding' refers to placing one element inside another. Thus one can take the phrase 'John's father' and place this inside the larger phrase 'John's father's uncle'. Both of these are embedding. The difference between embedding (which is common in languages that, according to researchers like Karlsson, have hierarchical structures without recursion) and recursion is simply that there is no bound on recursion, it can keep on going. So what Karlsson is saying is that in Standard Average European (SAE) languages he only found sentences like:

John said that Bill said that Bob is nice, or even
John said that Bill said that Mary said that Bob is nice, but never
  sentences like
*John said that Bill said that Mary said that Irving said that Bob is
  nice.*

The first sentence of this triplet has one level of embedding, the second has two and the third has three. But the claim is that standard European languages never allow more than two levels of embedding. Therefore, they have hierarchy (one element inside another), but they do not have recursion (one element inside another inside another ... ad infinitum).

According to Karlsson's research it does not appear that *any* SAE language is recursive. He acknowledges that one could argue that they are recursive in an abstract sense, or that they are generated by a recursive process. But such an approach doesn't seem to match the facts in his analysis. In other words, Karlsson claims, to use my terms, that all of these languages are G_2 languages. Karlsson's work is therefore interesting support from modern European languages that G_2 languages exist, as per the semiotic progression. Again, these are languages that are hierarchical but not recursive. To show beyond doubt that a language is recursive we need to show 'centre-embedding'. Other forms are subject to alternative analyses:

(a) Centre-embedding: 'A man that a woman that a child that a
    bird that I heard saw knows loves sugar.'
(b) Non-centre-embedding: 'John said that the woman loves
    sugar.'

What distinguishes (a) is that the subordinate clauses, 'that a woman that a child that a bird I heard saw knows' are surrounded by constituents of the main clause, 'A man loves sugar'. And 'that a child' is surrounded by bits of the clause it is embedded within 'a woman knows', and 'that a bird' is surrounded by parts of its 'matrix' clause 'a child saw' and so on. These kinds of clauses are rare, though, because they are so hard to understand. In fact, some claim that they exist only in the mind of the linguist, though I think that is too strong.

In (b), however, we have one sentence following another. 'John said' is followed by 'that the woman loves sugar'. Another possible analysis is that 'John said that. The woman loves sugar.' This analysis (proposed by philosopher Donald Davidson) makes English look more like Pirahã.[4]

The final type of language I am proposing in line with Peirce's ideas, G_3, must have both hierarchy and recursion. English is often claimed to be exactly this type of language, as previous examples show. As we saw earlier some linguists, such as Noam Chomsky, claim that *all* human languages are G_3 languages, in other words, that all languages have *both* hierarchy and recursion. He even claims that there could not be a human language without recursion. The idea is that recursion is what separates in his mind the communication systems of early humans and other animals from the language of *Homo sapiens*. An earlier human language without recursion would be a subhuman 'protolanguage' according to Chomsky.*

However, the fact remains that no language has been documented in which any sentence is endless. There may be theoretical reasons for claiming that recursion underwrites all modern human languages, but it simply does not match the facts of either modern or prehistoric languages or our understanding of the evolution of languages.

There are many examples in all languages that show non-direct connection between syntax and semantics. In discourses and conversations, moreover, the meanings are composed by speakers from disconnected sentences, partial sentences and so on. This supports the idea that the ability to compose the larger meanings of conversations from their parts (or 'constituents') is mediated by culture. As with G_1 languages and people suffering from aphasia, among many other cases, meaning is imputed to different utterances by loosely applying cultural and individual knowledge. Culture is always present in interpreting sentences,

---

*See Marc Hauser, Noam Chomsky and Tecumseh Fitch, 'The Faculty of Language: What Is It, Who Has It, How Did It Evolve?' *Science* 298, 2002: 1569–1579. Though the authors use the term 'recursion', they now claim that they do not mean recursion as understood by people doing research outside of Chomsky's Minimalism, but they actually intend Merge, a special kind of grammatic operation. This has caused tremendous confusion, though, ultimately, the issues do not change and Merge has been falsified in several modern grammars (see my *Language: The Cultural Tool*, among many others).

though, regardless of the type of grammar, G_1, G_2, or G_3. We know not exactly which of these three grammars, or all three, were spoken by *Homo erectus* communities, only that the simplest one, G_1, would have been and still is adequate as a language, without qualification.

It should ultimately be no surprise that *Homo erectus* was capable of language, nor that a G_1 language would have been capable of taking them across the open sea or leading them around the world. We are not the only animals that think, after all. And the more we can understand and appreciate what non-human animals are mentally capable of, the more we can respect our own *Homo erectus* ancestors. One example of how animals think is Carl Safina's book *Beyond Words: What Animals Think and Feel*. Safina makes a convincing case that animal communication far exceeds what researchers have commonly noticed in the past. And other researchers have shown that animals have emotions very similar to human emotions. And emotions are crucial in interpreting others and wanting to communicate with them, wanting to form a community. Still, although animals make use of indexes regularly and perhaps some make sense of icons (such as a dog barking at a television screen when other dogs are 'on' the screen), there is no evidence of animals using symbols in the wild.*

Among humans, however, as we have seen, there is evidence that both *Homo erectus* and *Homo neanderthalensis* used symbols. And with symbols plus linear order we have language. Adding duality of patterning to this mix, an easy couple of baby steps get us to ever more efficient languages. Thus possession of symbols, especially in the presence of evidence that culture existed, evidence strong for both *erectus* and *neanderthalensis*, indicates that it is highly likely that language was in use in their communities.

It is worth repeating that there is no need for a special concept of 'protolanguage'. All human languages are full-blown languages. None is inferior in any sense to any other. They each simply use one of three

---

* *This is not to say that animals could not use symbols.* I just am not aware of any well-supported or widely accepted claims that they can, in the wild. I certainly accept the idea that gorillas and other creatures have been taught to use symbols in the lab, and there are cases of animals, such as Koko the gorilla, using symbols after instruction.

strategies for their grammars, G_1, G_2, or G_3. Thus I find little use for this notion, given the theory of language evolution here.

The question of 'what evolved', therefore, eventually gets us back to two opposing views of the nature of language. One is Chomsky's, which Berkeley philosopher John Searle describes as follows:

The syntactical structures of human languages are the products of innate features of the human mind, and they have no significant connection with communication, though, of course, people do use them for, among other purposes, communication. The essential thing about languages, their defining trait, is their structure. The so-called 'bee language', for example, is not a language at all because it doesn't have the right structure and the fact that bees apparently use it to communicate is irrelevant. If human beings evolved to the point where they used syntactical forms to communicate that are quite unlike the forms we have now and would be beyond our present comprehension, then human beings would no longer have language, but something else.[5]

Searle concludes, 'It is important to emphasize how peculiar and eccentric Chomsky's overall approach to language is.'

A natural reply to this would be that what is one person's 'peculiar and eccentric' is another person's 'brilliantly original'. There is nothing wrong per se with swimming against the current. The best work often is eccentric and peculiar. But I want to argue that Chomsky's view of language evolution is to be questioned not simply because it is original, but because it is wrong. He has continued to double-down on this view for decades. In his recent book on language evolution with MIT professor of Computer Science Robert Berwick, Chomsky presents a theory of language evolution which furthers his sixty-year-old programme of linguistic theorising, the programme that Searle questions above. Chomsky's view was so novel and shocking in the 1950s that it was initially thought by many to have revolutionised linguistic theory and been the first shot fired in the 'cognitive revolution' that some date to a conference at MIT on 11 September 1956.

But Chomsky's linguistic theory was neither a linguistic revolution nor a cognitive one. In the 1930s Chomsky's predecessor, and I would say his inspiration, Leonard Bloomfield, along with Chomsky's PhD thesis supervisor Zellig Harris, developed a theory of language remarkably like Chomsky's, in the sense that structure rather than meaning was central and communication was considered secondary. And another predecessor of Chomsky's, Edward Sapir, had since the 1920s argued that psychology (what some today would call cognition) interacted with language structures and meanings in profound ways. In spite of these influences, Chomsky has staked out his claims to originality clearly over the years, reiterating them in his new work on evolution, namely that 'language' is a computational system, not a communication system.*

So for Chomsky there is no language without recursion. But the evidence from evolution and modern languages paints a different picture. According to the evidence, recursion would have begun to appear in language, as we saw earlier, via gestures, prosodies and their contributions to the decomposition of holophrastic utterances.

As speech sounds produced auditory symbols (words and phrases) these symbols would have been used in larger strings of symbols. Gestures and intonation, whether precisely aligned or only perceived to be aligned with specific parts of utterances, would have led to a decomposition of symbols. Other symbols could have been derived from utterances that had little internal structure initially, but were then likewise broken down via gestures, intonation and so on.

The bottom line is that recursion is secondary to communication and that the fundamental human grammar that made possible the first human languages was a G_1 grammar.†

Chomsky's grammar-first theory is disconnected from the data of human evolution and the cultural evidence for the appearance of advanced communication. It ignores Darwinian gradual evolution,

---

*More technically, language is nothing more nor less than a set of endocentric, binary structures created by a single operation, Merge, and only secondarily used for storytelling, conversation, sociolinguistic interactions and so forth.

†Recursion simply allows speakers to pack more information into single utterances. So while 'You commit the crime. You do the time. You should not whine.' are three separate utterances we can say the same thing in one utterance using recursion: 'If you commit a crime and do the time, you should not whine.'

having nothing to say about the evolution of icons, symbols, gestures, languages with linear grammars and so on, in favour of a genetic saltation endowing humans with a sudden ability to do recursion. Again, according to this theory, communication is not the principal function of language. While all creatures communicate in one way or another, only humans have anything remotely like language because only humans have structure-dependent rules.*

---

*This is circular in the sense that Chomsky takes a feature that only humans are known to have, structure-dependency, and claims that, because this defines language, only humans have language.

10

Talking with the Hands

Small gestures can have a big impact.

Julianna Margulies

JUST AS TACIT CULTURAL KNOWLEDGE shapes grammars, it is also important in each of the latter's components – words, gestures, phonology, syntax, discourse and conversation. However, for many linguists and anthropologists, gestures are often omitted from discussion, judged too quickly to be secondary accoutrements of speech, a separate, independent facet of human behaviour. But researchers from various theoretical perspectives have shown, to the contrary, that an intimate set of connections exists between hand movements, linguistic structure and cognition, held together by tacit, cultural knowledge. An analysis of the symbiosis between the hands, the mouth and the brain and how these evolved has to round out any theory of language evolution.[1]

Additionally, some have argued that this all shows something more, namely that the highly language-specific components of gestures are innate. This research, pioneered in the work of Susan Goldin-Meadow, examines the 'spontaneous emergence' of hand movements in children who have otherwise no access to linguistic input, as in the deaf children of hearing, non-signing parents. She calls these gestures 'homesigns' and the gestural systems she studies might indeed be crucial to our quest to tease apart native vs cultural or a priori vs a posteriori perspectives on the origins of (some) dark matter.

To understand the role of gestures in language, it is important to grasp how they work together with intonation, grammar and meaning. It is possible to get some idea of how these different abilities combine by focusing on gestures and intonation as 'highlighters', helping hearers

pick out new or important information from old information that the
speaker assumes to be shared with the hearer. One way to do this is
to examine the evolving research into gestures and human language,
from the ancients through the current and very important research
of contemporary scientists. Without understanding gestures there is
no understanding of grammar, the evolution of language, or the use
of language. Gestures are vital for a fuller understanding of language,
its origins and its broader role in human culture, communication and
cognition.

Language is holistic and multimodal. Whatever a language's
grammar is like, language engages the whole person – intellect, emotions, hands, mouth, tongue, brain. And language likewise requires
access to cultural information and unspoken knowledge, as we produce
sounds, gestures, pitch patterns, facial expressions, body movements
and postures all together as different aspects of language. I want to begin
here with an overview of the functions and forms of gestures in the
world's languages, including most likely the language(s) of early *Homo*
species. Gestures can be complex or simple. But they can be learned.

The gestures that accompany all human speech reveal an intersection
of culture, individual experience, intentionality and other components
of 'dark matter', or tacit knowledge. There are two kinds of knowledge
of human grammars, as there are of most things: the static and the
dynamic. These are quite possibly related to declarative and procedural
memories, but they do seem a bit different. Static knowledge is a list
of the things we know. Rules for telling stories are static knowledge.
Dynamic knowledge, however, is understanding that things change
and knowing how to adapt to changes in real time. If static knowledge is knowing the rules for telling a story, dynamic knowledge is
telling the story. Gestures are crucial components of our multimodal
languages. They are themselves intricate in structure, meaning and
use. Contemporary research makes it clear that gesticulations are as
analytically challenging and as intricate in design and function as any
other part of language. But, to reiterate, these are not simply add-ons to
language. There cannot be a language without gestures. Most of these
are used unconsciously and employ tacit knowledge. They are shaped
by the needs of the language they enhance and the cultures from which
they emerge.

Kenneth Pike saw gestures as evidence for the idea that language should be studied in relation to a unified theory of human behaviour:

> In a certain party game people start by singing a stanza which begins Under the spreading chestnut tree ... Then they repeat the stanza to the same tune but replace the word spreading with a quick gesture in which the arms are extended rapidly outward, leaving a vocal silence during the length of time which would otherwise have been filled by singing. On the next repetition the word spreading gives place to the gesture, as before, and in addition the syllable chest is omitted and the gap is filled with a gesture of thumping the chest. On the succeeding repetition the head is slapped instead of the syllable nut being uttered ... Finally, after further repetitions and replacements, there may be left only a few connecting words like the, and a sequence of gestures performed in unison to the original timing of the song.[2]

Pike concludes from this example that gestures can replace speech. Later researchers, however, have shown that the gestures he refers to are a very limited type out of several kinds that are possible. Language is just a form of behaviour, as gestures also are. Still, Pike's basic point is valid – language and its components are human behaviour guided by individual psychology and culture, dark matter of the mind.

All human behaviour, including language, is the working out of intentions, what our minds are directed towards. Language is the best tool for communicating those intentions. Communication is a cooperative behaviour. It follows cultural principles of interaction.

Pike raised another question: why don't people mix gestures or other noises with speech sounds in their grammars? Why is it that only sounds made by the mouth can be used in syllables and speech more generally? Why couldn't a word like 'slap' be [sla#], where [#] would represent the sound of someone slapping their chest? It may sound easy, but really this is not a possible word or syllable in anyone's language. As a beginning linguistics student, I thought this question was interesting but did not appreciate adequately the degree to which it impinges on the understanding of language.

Gestures aim towards what linguists and philosophers call

'perlocutionary effects', the effects that a speaker intends her language to have on a hearer. Speakers use highlighters in order to help the listener use or react to the information in the way the speaker hopes they will.

To more fully illustrate the need for a single theory of culture and language, indeed all human behaviour, one might contemplate a scene like the following. Two men are watching other men move some heavy furniture down the stairs in their apartment building. One man passing on the stairway landing is huffing and puffing and concentrating solely on his heavy load. His wallet is hanging loosely from his back pocket, about to fall out. He clearly wouldn't notice if someone relieved him of this burden. The first observer looks at the second observer with raised eyebrows, looking at the wallet. The second one sees him and simply shakes his head to indicate 'No'. What happened here? Is this language? It is a form of communication that is parallel to language. Certainly shared culture and conventions are necessary for this kind of exchange. Just about anything two members of a culture wish to exploit can be used to communicate.

There is a broad popular interest in gestures, though people often fail to recognise how fundamental they are to language. They formed the basis of a 2013 article in the *New York Times* by Rachel Donadio titled 'When Italians Chat, Hands and Fingers Do the Talking'.[3] Italians do indeed stand out gesturally, but so do we all. Even in the seventeenth century northern European Protestants disapproved of Italians' 'flamboyant' hand movements. But the first person to study Italians', or any other language's, gestures from a modern scientific perspective was David Efron, a student of twentieth-century pioneer in anthropology and linguistics Frans Boas. Efron wrote the earliest modern anthropological linguistic study on cultural differences in gestures more than seventy years ago. He focused on the gestures of recent Italian and Jewish immigrants and later compared those with the gestures of second- and third-generation immigrants.

Efron's study, *Gesture, Race and Culture*, was simultaneously a reaction against Nazi views of the racial bases of cognitive processes, a development of a model for recording and discussing gestures and an exploration of the effects of culture on gesture. The core of Efron's contribution is his description of the gestures of unassimilated southern Italians and east European Jews ('traditional' Italians and Jews), recently emigrated to the United States and mainly living in New York City (though some of his subjects also came from the Adirondacks, Saratoga and the Catskills). According to Efron, Italians use gestures to signal and support content. For example, a 'deep' valley, a 'tall' man, 'no way'. The Jewish immigrants of Efron's study, on the other hand, use gestures as logical connectives, that is to indicate changes of scene, logical divisions of a story and so on. These uses of gesture underscore the fact that language is triple-patterned (symbols, structure and high-lighters, such as gestures and pitch) and shaped by culture.

Efron wanted to know two things about gestures. First, are there standardised group differences in gestures between Italian and Jewish immigrants? Second, he wanted to understand how gestures change when an immigrant is socially assimilated. He discovered a strong cultural effect. An 'Americanisation' of gestures occurred in each group over time. The initially strong differences between Jewish vs Italian immigrants grew less pronounced until they were identical to any other citizen of the United States.

Since his was a pre-video-camera era, Efron contracted an artist to help him, Stuyvesant Van Veen. Efron was the first to come up with an effective methodology for studying and recording gestures, as well as a language for describing them. Although later parts of the book orthogonally attacked Nazi-science, the book was a breakthrough. Efron's work, though pioneering, emerged from a long tradition.

Aristotle discouraged the overuse of gestures in speech as manipulative and unbecoming, while Cicero argued that the use of gestures was important in oratory and encouraged their education. In the first century, Marcus Fabius Quintilianus actually received a government grant for a book-length study of gesture. For Quintilian and most of the other classical writers, however, gesture was not limited to the hands but included general orientation of the body and facial expressions, so-called 'body language'. In this they were correct. These early explorers of gesture in human languages discovered that communication is holistic and multimodal.

The Renaissance rediscovered the work of Cicero and other classical scholars, sparking European interest in the relationship between gesture and rhetoric. The first book in English on gesture was John Bulwer's *Chirologia: or the Naturall Language of the Hand* in 1644.

By the eighteenth century researchers on gesture began to wonder whether gestures might have been the original source for language. This idea is echoed by several modern researchers, but it is one that should be discouraged. Gestures that can serve in place of speech, such as sign languages or mimes, in fact repel speech, as University of Chicago psychologist David McNeill has shown. They replace it. They are not replaced by speech, which would have to be the evolutionary progression if gestures came first.

Interest in understanding the significance and role of gestures in human language and psychology diminished tremendously, however, in the late nineteenth and early to mid-twentieth centuries. There were several reasons for this decline. First, psychology was more interested in the unconscious than in conscious thinking during this period and it was thought, erroneously, that gestures were completely under conscious control. Gesture studies also dwindled because linguists became more interested in grammar, narrowly defined by some so as to exclude gestures. The interest in the messy multimodality of language had waned. Another factor leading to the decline of gesture studies was that linguistic methods of the day were still not up to the task of studying gestures scientifically. Efron's work was extremely hard to do and wasn't amenable to widespread duplication, at least as many perceived it at the time. Not everyone can afford an artist.

Linguist Edward Sapir was different. He saw language and culture as two sides of a coin. Therefore, his view of gestures was similar to those of current research. As Sapir said, 'the unwritten code of gestured messages and responses is the anonymous work of an elaborate social structure'. By 'anonymous' Sapir meant tacit knowledge or dark matter.

This raises the fundamental and obvious question: what are gestures? Are sign languages gestures? Is a mime gesture? Are signals such as the 'OK' sign with the thumb and forefinger or 'the bird' with an upraised middle finger gestures? Yes to all the above. Some researchers, such as David McNeill and Adam Kendon, classify all these different forms along a 'gesture continuum' that looks at gestures in terms of their dimensions and their relationship to grammar and language (Figure 32).

Gesticulation, the most basic element of the continuum, is the core of the theory of gestures. It involves gestures that intersect grammatical

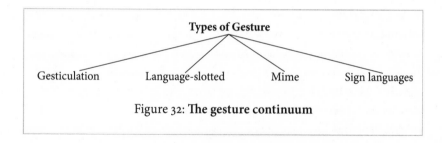

Figure 32: **The gesture continuum**

structures at the place where gestures and pitch and speech coincide. Gesticulation is, in fact, what most theories of gesture are about. Some gestures are not conventional – they may vary widely and have no societally fixed form (though they are culturally influenced). Gestures that can replace a word, as in Pike's 'chestnut tree' language game, are 'language-slotted'. These can also be seen if you were to tell someone, 'He (use of foot in a kicking motion) the ball,' where the gesture replaces the verb 'kicked', or, 'She (use of open hand across your face) me,' (for 'She slapped me'). These gestures occupy the positions in sentences usually taken by words. They are special gestures. These gestures are improvised and used to produce particular effects according to the type of story being told. Fascinatingly, these language-slotted gestures are a window into speakers' knowledge of their grammars. One cannot use gestures unless one knows how words, grammar, pitch and the rest fit together.

Our hand movements can also simulate an object or an action without speech. When they do this we are using mime, which follows only limited social conventions. Such forms vary widely. Just play a game of charades with a group of friends to see that. Conventionalised gestures can also function as 'signs' in their own right. As I mentioned, two common emblems in American culture are the forefinger and the thumb rounding and touching at their tips to form the 'OK' sign and 'the bird', the upraised, solitary middle finger.

These are all distinct from sign languages. These are full-blown languages. They have all the features of a spoken language, such as words, sentences, stories and even their own gestural and intonational highlighters. These are expressed in sign languages by different kinds of body and hand movements and facial expressions. In our discussion of language evolution, it is very important to keep in mind the most salient feature of gesture-based languages. That is that sign languages neither enhance nor interact with spoken language. In fact, *sign languages repel speech*, to use one of McNeill's phrases. This is why many researchers believe that spoken languages did not and could not have begun as sign languages.

Now let's move to the crux of gesture's relevance for language evolution. The bedrock concept here, developed in McNeill's research, is called the 'growth point'. The growth point is the moment in an utterance where gesture and speech coincide. It is where four things happen. First, speech and gesture synchronise, each communicating different yet related information simultaneously.

The growth point is further described as that point where gesture and speech are redundant, each saying something similar in different ways, as in Figure 32. The gesture highlights a newsworthy item from the background of the rest of the conversation, again as in Figure 32. Intonation, it should be mentioned, is also active at the growth point and other places in what is being said. Third, at the growth point gesture and speech communicate a psychologically unified idea. In Figure 33, the gesture for 'up' occurs simultaneously with the word for 'up'.

In short, gesture studies leave us with no alternative but to see language not as a memorised set of grammar rules but as a process of communication. Language is not static, only following rigid grammatical specifications of form and meaning, but it is dynamic, bringing together pitch, gestures, speech and grammar on the fly for effective communication. Language is manufactured by speakers in real time, following their tacit knowledge of themselves and their culture. Gestures are actions and processes par excellence. The boundaries between gestures are clear, being the intervals between successive movements of the limbs, according to McNeill. Like all symbols, gestures too can be decomposed into parts. I won't go into these here except to say that this all means that gestures, intonation and speech become a multimodal, holistic system, requiring a *Homo* brain to orchestrate their cooperative action.

Another crucial component of the dynamic theory of language and gestures that McNeill develops is the *catchment*. This is a bit technical, but it is essential to understanding how gesture facilitates

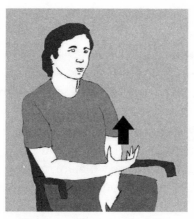

Figure 33: **The growth point**

communication and thus the potential role of gesture at the beginning of language. A catchment indicates that two temporally discontinuous portions of a discourse go together – repeating the same gesture indicates that the points with such gestures form a unit. In essence a catchment is a way to mark continuity in the discourse through gestures. McNeill says:

> [A] catchment is recognized when one or more gesture features occur in at least two (not necessarily consecutive) gestures. The logic is that recurrent images suggest a common discourse theme and a discourse theme will produce gestures with recurring features ... A catchment is a kind of thread of visuospatial imagery that runs through a discourse to reveal the larger discourse units that encompass the otherwise separate parts.[4]

Assume that while speaking you use an open hand, turned upward with the fingers also pointed upward, whenever you repeat the theme about a friend wanting something from you. The gesture then becomes associated with that theme, highlighting the theme thereby and helping your hearer follow the organisation of your remarks more easily.

In other words, through the catchment, gestures enable speakers to arrange sentences and their parts for use in stories and conversations. Without gestures there could be no language.

Various experiments have been developed that illustrate an

'unbreakable bond' between speech and gestures. One of the more famous experiments is called delayed auditory feedback. For this test the subject wears headphones and hears parts of their speech on roughly a 0.2 second delay, close to the length of a standard syllable in English. This produces an auditory stuttering effect. The speaker tries to adjust by slowing down. The reduced rate of speech offers no help, however, because the feedback is also slowed down. The speaker then simplifies their grammar. On top of this, the gestures produced by the speaker become more robust, more frequent, in effect trying to take more of the communication task upon themselves. But what is truly remarkable is that the gestures stay synchronised with the speech no matter what. Or, as McNeill puts it, the gestures 'do not lose synchrony with speech'. This means that gestures are tied to speech not by some internal counting process, but by the intent and meaning of the speaker. The speaker adjusts the gestures and speech harmoniously in order to highlight the content being expressed.

Other experiments also illustrate clearly the tight connection between speech and gestures in normal talk. One experiment involves a subject referred to as 'IW'. At age nineteen, IW suddenly lost all sense of touch and proprioception below the neck due to an infection. Experiments show that IW is unable to control his hand movements unless he can see his hands (if he cannot see them, such as when they are below the table he is seated at, then he cannot control them). What is fascinating is that IW, when speaking, uses gestures that are well coordinated, unplanned and closely connected to his speech as though he had no disability at all. The case of IW provides evidence that speech gestures are different from other uses of the hands, even other gesturing uses of the hands. Some suggest that this connection is innate. But we know too little about the connection of gestures and speech in the brain or the physiological history of IW to conclude this. In any case, however, this coordination comes about, gestures in speech are very unlike the use of our hands in any other task.

One final observation to underscore the special relationship between gestures and speech: even the blind use gestures.* This shows

---

*To say that this also means that we use gestures without learning them, as McNeill seems to suggest, is unwarranted.

that gestures are a vital constituent of normal speech. The blind's use of gestures has yet another lesson for us. Since the blind cannot have observed gestures in their speech community, their gestures will not match up exactly to those of the local sighted culture. And yet this very fact shows that gestures are part of communication and that language is holistic. We use as much of our bodies as we are able when we are communicatively engaged. We 'feel' what we are saying in our limbs and faces and so on.

The connection between gestures and speech is also culturally malleable. Field researchers have demonstrated that the Arrernte people of Australia regularly perform gestures after the speech. I believe that the reason for this is simple. The Arrernte simply prefer gestures to follow speech. The lack of synchrony between gestures and speech is simply a cultural choice, a cultural value. Gestures for the Arrernte could then be interpreted similarly to the Turkana people of Kenya, in which gestures function to echo and reinforce speech.

Were gestures also important for *Homo erectus*? I believe so, based, once again, on the work of David McNeill. He introduces the term 'equiprimordiality', by which he means that gestures and speech were equally and simultaneously present in the evolution of language. There never would have been nor could have been language without gestures.

If this is correct, claims McNeill, then 'speech and gesture had to evolve together'. 'There could not have been gesture-first *or* speech-first.' This follows because of my concept of triality of patterning. You cannot have language without grammar, meaning and highlighters. By the same token, there could never have been intonation without language or language without intonation.

Once this initial hurdle of how gestures become meaningful for humans is overcome, the evolutionary story of the connection between gesture and speech may be addressed. McNeill's theory hypothesises that early speech by the first speakers and human infants was 'holophrastic'. That is, in these early utterances there are no 'parts', only a whole. To return to an earlier example, say that the first utterance by an *erectus* was 'Shamalamadingdong!' as he saw a sabre-toothed cat run by him only a hundred yards away. He was in all likelihood gesticulating, screaming and engaging his entire body to communicate what he had seen, unless he was frozen with fear. His body and head would have been directed towards the cat. Later, perhaps he recreated this scene, using slightly different gestures and intonation (he is calm now). The first time perhaps he uttered SHAMALAmadingDONG, with hand movements on shama and dong. The next time perhaps his intonation fell on shamalamaDINGdong. Perhaps his gestures remained over 'shama' and 'dong' or, more likely, they were more closely linked to any change in his intonation. It is now possible that *erectus* has inadvertently taken a holophrastic – single unit – utterance and transformed it into a construction with individual parts. And this is how McNeill proposes that grammar began to emerge.

As gestures and speech become synchronised, gestures can then show one of two characteristics. They either represent the viewpoint of the observer – the viewpoint of the speaker – or they represent the viewpoint of the person being talked about. And with these different viewpoints, different ways of highlighting content and attributing ownership of content, we lay the groundwork for distinctions among utterances such as questions, statements, quotes and other kinds of speech acts.

McNeill gives an example of one person retelling what they saw in a cartoon of Sylvester the cat and Tweety Bird. When their hand movements are meant to duplicate or stand for Sylvester's movements, then their perspective is Sylvester's. But when their hand movements indicate their own perspective, then the perspective is also their own.*

Intentionality – being directed at something – is also a prerequisite to having a language. And intentionality is shown not only in speech but also in gestures and other actions. We see it in anxiety, tail-pointing in canines and in focused attention across all species. One reason that gestures are used is because intentional actions engage the entire body. The orientation of our eyes, body, hands and so on varies according to where we are focusing our attention. This holistic nature of expressing intentions seems to be a very low-level biological fact that is exploited by communication. The fact is, 'animals use as much of their bodily

---

*As stated, many researchers have speculated that gestures might have preceded speech in the evolution of human language. It is possible that gestures were first in some way, preceding language proper. Though I should have first expected yells along with gestures. Even McNeill does not disagree entirely with this position.

resources as necessary to get their message across'. If we are on the right track, though, gestures could not have been the initial form of language. They would have occurred simultaneously with intonation and vocalisation. This is not to say that prelinguistic creatures cannot express intentionality by pointing or gesturing in some way. It does mean that real linguistic communication must have always included both gestures and speech. There are a few additional reasons for this judgement.

First, speech did not replace gesture. Gestures and speech form an integrated system. The gesture-first origin of language predicts a mismatch between gesture and speech, since they would be separate systems. But in reality they are synchronous (they match in time) and parts of a single whole (a gesture plus intonation plus speech coordinated in a single utterance). Further, people regularly switch between gestures and speech. Why, if speech evolved from gestures, would the two still have this give-and-take relationship? Finally, if the gesture-first hypothesis is correct, then why, aside from languages of the deaf, is gesture never the primary 'channel' or mode of communication for any language in the world?

Intonation was alluded to earlier when discussing 'Yesterday, what did John give to Mary in the library?' Whenever we speak we also produce a 'melody' over our words. If an example of the importance of intonation is desired, one need only think about how artificial a car's GPS sounds when it's giving directions. Although computer scientists long ago learned that speech requires intonation, they still have not produced a computer that can use or interpret intonation well. Intonation, gestures and speech are built upon a stable grammar. The only gestures that provide stability are the conventionalised and grammaticised gestures in sign languages. In this case again, however, gestures are either used instead of or to supplant speech.

What is crucial is that gestures co-evolved with speech. If sign language, language-slotted gestures or mimes had preceded speech, then there would have been no functional need for speech to develop. The gesture-first idea stakes out an untenable position. We had a well-functioning gestural communication but replaced it wholesale with speech. And some gestures, such as mimes, actually are incompatible with speech.

This might seem to be contradicted by the earlier example from

Kenneth Pike that apparently shows that gestures can substitute for speech. But the gestures Pike discusses are language-slotted gestures, a distinct kind of gesture parasitic on speech, not the type of gesture to function in place of speech. On the other hand, Pike's example suggests another question, namely whether there could be 'gesture-slotted speech' corresponding to speech-slotted gestures. This would be a case of an output in which speech substitutes for what would usually be expressed by gestures. If speech evolved from gestures, after all, this is how it would have come about. And gesture-slotted speech is not hard to imagine. For example, consider someone bilingual in American Sign Language and English substituting a spoken word for each sign, one-by-one, in front of an audience. Yet such an event would not really exemplify gesture-slotted language, since it would be a translation between two independent languages, not speech replacing gestures in a single language. This is important for our point for a couple of reasons. The obviously utilitarian nature of hand signs offers us a clear route to understanding their origin and spread. And the fact that everyone seems to use gestures in all languages and cultures of the world is supportive of the Aristotelian view of knowledge as learned, over the Platonic conception of knowledge as always present. This follows because it shows that the usefulness of gestures is the key to their universality. When a behaviour is an obvious solution to a problem, there is no need to assume that it is innate. The problem alone guarantees that the behaviour will arise if the mind is intelligent enough. This principle of usefulness explains most supposedly universal characteristics of language that are often proposed to be innate. In other words, their utility explains their ubiquity.

As they stabilise by conventionalisation, gestures become sign languages. But sign languages are formed when gestures replace all speech functions. The idea that speech develops from gestures thus makes little sense either functionally or logically. The 'gesture-first' theory gets the direction of evolution backwards.

However, in spite of my overall positive view of McNeill's reasoning about the absence of gesture-first languages, there seems to be something missing. If he were correct in his additional assertion or speculation that two now-extinct species of hominin had used either a gesture-first or a gesture-only language, and that this is the first stage in the development of modern language, then why would it be so surprising to think that *Homo sapiens* had also used gesture-first initially? I see no reason to believe that the path to language would have been different for any hominin species. In fact, I doubt seriously that pre-*sapiens* species of *Homo* would have followed a different path, since there are significant advantages to vocal vs gesture communication.

There are still other types of gesture important in human communication. These include iconic gestures, metaphoric gestures and beats. Each reveals a distinct facet of the gesture–speech relationship and its relationship to cognition and culture. There is no need to discuss these here, other than to mention them as evidence for the complexity of the relationship between gestures and speech and that they each contribute to our progress along the semiotic progression.

Nevertheless, we have yet to see anything in grammar, gestures, or other aspects of language that would lead us to believe that anything needs to be attributed to the genome of *Homo sapiens* that is specific to language. Cultural learning, statistical learning and individual apperceptional learning complemented with human episodic memory seem up to the task. The literature is rife with claims to the contrary, namely that there are phenomena that can be explained only if language is acquired at least partially based on language-specific biases in the newborn learner.

It is claimed that there is a spontaneous emergence of language features in communities that are claimed to otherwise lack language, such as Nicaraguan Sign Language and Al-Sayyid Bedouin Sign Language. These languages are purported to come into existence suddenly as a population needs language but otherwise lacks one. The problem with this claim is that all of these languages begin with very simple structures and then become more complex over time with more social interaction. Often it takes at least three generations to develop a complexity roughly equal to better-established languages. But this is just what we would expect if they were not derived from innate knowledge but invented and improved over time as they were learned by subsequent generations. For this reason, even if such examples provided some evidence for an innate predisposition to language, the knowledge would be very limited.*

---

*More importantly, what marks the work of Goldin-Meadow and many others is

Susan Goldin-Meadow's work argues that homesigners develop symbols for objects, principles for ordering these and constituents of distinct gestures. She also suggests that these newly minted gestures can fill slots in larger sentence-like structures, structures represented by the kind of tree diagrams we saw earlier. She also discusses a number of other characteristics of homesigns. Her conclusion is that all of this knowledge must be innate or how else could it appear so quickly among a group of speakers?

But none of these characteristics is specific to language. Indexes and icons, in all probability early forms of gestures, are used in one way or another by several species. There is no reason to believe that homesigns cannot be learned easily by humans. In fact, on one interpretation, that is all that Goldin-Meadow's results on symbols show us, namely that children readily learn and adopt symbols. The object is a form with a meaning. As the child learns the object and desires to communicate, then – perhaps the most striking characteristic of our species – whether due to an interactional instinct or an emotional urge, the child will represent the object and the meaning of the object in the particular culture comes along for the ride. Children participate in their parents' lives, even if without language, and try to communicate, as Helen Keller's remarkable odyssey shows us. With an ability to see or hear or feel, the child can receive input from the environment, from its caregivers, and in fact will do so with most caregivers and in most environments. Learning the use of the object, the salience of the object to its parents and environment, it is unsurprising that children communicate about objects, as most other species (at least mammal species) do. Whole objects, as perceivable in a particular space in time, are most salient and learned relatively easily by dogs, humans and other creatures. Humans try to represent their objects, unlike other animals, because humans strive to communicate.

The fact that some features of the objects stand out to children is likewise unsurprising, though the particular reason that shape and

---

what I consider to be an over-charitable interpretation of the linguistic aspects of the signs and a less charitable view of the cultural input the child receives as well as the nature of the task the child is facing. Absent a serious consideration of either the task or the input, such claims are severely weakened.

size win out over many other features, if Goldin-Meadow is correct, is unclear. She ascribes it to the child's native endowment. But I would suggest looking first at the way that objects are used, presented, structured and valued in the examples of the child's caregivers. Furniture, dishes, houses, tools and so on are far more easily arranged and far more prevalent in the environment of US caregivers' salient objects than other features. At least that could be tested and there is no suggestion that any such tests were contemplated.

With regard to the claim that homesigners' speech is organised hierarchically, there are two caveats. First, structure vs simple juxtapositioning of words like beads on a string are very difficult to distinguish in practice. Are three objects related as in diagram (a) or (b) of Figure 34? Either might be the case, and the reasons for choosing one over the other are highly theoretical. For example, in Pirahã utterances, we might say, 'The man is here. He is tall.' Or, 'I spoke. You are coming.' And these could be interpreted as 'The man who is tall is here,' or 'I said that you are coming.' But the analysis is quite possibly much simpler, with the syntax lacking hierarchical structure.

In none of Goldin-Meadow's examples purporting to show hierarchical structure in homesigners' utterances is there convincing evidence for structures like (b). The second caveat is that some configurations provide a natural solution to presenting information, independent of language, and thus if one finds them in some languages, this is not evidence that there is an innate linguistic bias. Again, if a structure provides a useful solution to communicating information then nothing else needs to be said about why it is found in languages around the world. As information demands grow due to increasing societal complexity, hierarchy is the most efficient solution to information organisation, across many domains. Computers, atoms, universes and many other complex objects of nature are organised this way. It is a naturally occurring and observable solution. In fact, for any action that involves ordering, such that 'you must do x before you do y', there is structure. Such solutions are used in automobiles, canine behaviours and computer filing systems. There is absolutely nothing special about them when they also appear in language.

The ordering that homesigners are claimed to impose on their structures is mundane. First, they have no alternative but to put their symbols

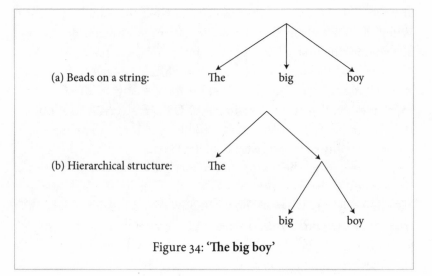

(a) Beads on a string:

(b) Hierarchical structure:

Figure 34: 'The big boy'

in some order. And since the main ingredients of any utterance are the thing being reported about and what happened to it, sentences tend to be organised in terms of topic and comment. The topic of a sentence (as opposed to the topic of a larger story) is the old information that is either being discussed or the information that the speaker assumes that the hearer will know. The comment is the new information about the topic. Very often, but not always, the topic is the same as the subject and the comment is the same as the predicate or verb phrase:

'John is a nice guy.'

Here the old information is 'John'. The speaker offers nothing more than a name and assumes that the hearer knows who is being mentioned. The new information is 'is a nice guy'. In other words, the speaker is saying something that he or she thinks might be new information for the hearer about 'John'. A paraphrase might be something like, 'I know you know John, but you might not know that he is a nice guy.'

The topic in most languages of the world precedes the comment. In other words, languages prefer to begin their sentences with shared or old information before giving new information. This order may aid our short-term memory. Within the comment, where the new information is placed, a large number of languages prefer to place the object before the verb. So if someone somewhere is eating a fruit, this can either be described as *John fruit eats* (majority of languages) or *John eats fruit* (English and many other languages). In the Subject Object Verb group are languages like German, Japanese and Pirahã. Languages like French and English, on the other hand, use the order Subject Verb Object. In fact, English used to belong to the former group (it is closely related to German), but after the Norman invasion of 1066 English switched to the French order of Subject Verb Object.

It is claimed occasionally that word orders may be heavily influenced by different communication strategies for dealing with problems in the communicative process. One problem is noise corrupting the signal. As people talk they can be distracted by background noise, children tugging on them, animals in the distance and new people arriving during the conversation. Therefore, linguistic strategies must be able to overcome the effect of 'noise' on the hearer's ability to perceive the speaker's words. These researchers claim that this is why Subject Object Verb order is most common – it helps avoid confusing the subject and the object and avoids confusing which is topic and which is comment, by having the phrase before the verb 'to be' in the comment section. On the other hand, since there are thousands of languages that use different orders, such as Subject Verb Object, Object Verb Subject, Object Subject Verb, Verb Subject Object and Verb Object Subject, no one of these orders is far superior to the others. The order that a language adopts results from cultural pressures that affect the history of a particular society. This is why that English changed in its history from Subject Object Verb to Subject Verb Object, as a result of the influence of the French spoken by the Norman conquerors.

Homesigning follows largely the same principles. The basic problem that interlocutors must solve is keeping novel information distinct from already known information. Thus the fact that homesigners follow a common order is simply not a big deal. It is just the way that communication works most efficiently.

Nor should it be surprising that once a basic sequence of words is made conventional, it is easier for speakers to remain consistent with that choice than to use different strategies in different parts of the language. Therefore, if your language chooses Subject + Object + Verb, it will, ceteris paribus, also choose to put possessors before nouns (as in 'John's book' vs 'Book John's'). This is because, just as the verb is the semantic core or nucleus of the sentence, the possessed noun is the nucleus of its own noun phrase. The general rule in such a language would be 'put the nucleus last'. Yet because this decision is based only on efficiency, and because languages often do things inefficiently, this type of rule is frequently violated all over the world and we observe a good deal of variation in word order principles in different languages for this reason. Nor is there anything specific in the genetic endowment of *Homo sapiens* that would be expected to be linked to information structure. Topic-comment is a natural communicative arrangement. But many who talk about the implications of homesigns, supposedly revealing innate linguistic capacity, neglect to discuss the way that information should ordered for most effective communication. Failure to refer to this, however, misses the most natural explanation for the facts about homesigns and appeals instead to the highly implausible idea that language is innate.

Homesigning clearly illustrates the desire of all members of *sapiens* to communicate. And it shows solutions to this communication problem can be straightforward and easy to 'invent' and understand. Unfortunately for the claims of homesign researchers that language is innate, there is not even a convincing analysis of the facts of the grammar of these languages. But evidence suggests that gestures are sufficiently motivated by communicational needs that sign languages and gestures of all kinds simply emerge because they are so useful. *Utility explains ubiquity.*

11
Just Good Enough

Le mieux est l'ennemi du bien.

Voltaire

Not everything worth doing is worth doing well.

Kenneth L. Pike

NOBEL-PRIZE-WINNING economist Herbert Simon introduced the concept of 'satisficing' into the science of problem solving. His point was that the solutions preferred by business, throughout human endeavours and by the mind itself are not usually the best ones but the good-enough ones – the ones that 'satisfice' the need rather than perfectly satisfy it. The same principle applies to evolution. With respect to language, this means that human grammars and sound systems are not required to be optimal – in fact, they never are. Language gets the job done just well enough, never perfectly. Herbert Simon has echoed Voltaire, claiming: 'The good is the eternal enemy of the best.'

It is this aspect of language that so strongly supports the idea that it is an ancient invention, tinkered with throughout human history. There are basic strategies of communication that usually work, but frequently fail – such as when one omits information that is assumed to be shared by one's interlocutor. Or communication can fail due to the failure to remember or the very lack of a word or, more importantly, a sentence, to translate a concept in one culture or person or language into one's native language. Human languages leak. They are not mathematical, perfectly logical codes.

If someone yells, 'Stop the car at the stop sign!' they assume that the person they are instructing knows what 'stop' means, how to drive

a 'car', what a stop sign is and what it means to stop at a stop sign as
opposed to stopping on the edge of a cliff. (One may stop at a sign
with the front wheels slightly beyond the sign. That wouldn't work out
well at a cliff.) Those assumptions are just built into the choice of the
words. I imagine that most readers of this book, as members of a largely
uniform driving culture, know that one has not 'stopped at the stop
sign' if they stop 200 yards in front of it or ten feet in front of it or even
if it is level with the rear seats. Proper stoppage at the signage brings
the vehicle to a complete stop roughly 1–5 feet before the front of the car
passes the stop sign. This is part cultural knowledge, part lexical (word)
knowledge. 'Common' sense is just experience and acquired cultural
information.

For a more involved example, consider a 2016 opinion piece published in the *New York Times* on the purported hacking of a presidential
candidate's emails by Russia. Most of the values embedded in this
opinion piece are easy to discover, though some are worth pointing out
for discussion because one often reads over such unspoken information
without noticing it. A few comments are offered, in square brackets
in order to reveal unspoken cultural and contextual information and
values that would be expected to be implicitly understood by every culturally North American reader. Imagine the potential effect of seeing
such editorials in a newspaper one takes on a regular basis over time,
reading them passively, perhaps, while sipping coffee at the breakfast
table or on the train or bus to work. Their effect could be largely sub-
liminal, reinforced with every story that assumes similar values, which
is the case of particular newspapers, with the *New York Times* usually
speaking for liberalism.

The Real Plot Against America
Timothy Egan – July 29, 2016

In retrospect [By beginning with a reference looking back, the author
appeals to the reader's assumed knowledge of what is to come],
it [Because the pronoun 'it' is used, the author tells the reader: 'I
assume you know this' – that which will become clear] *worked out
much better than planned.* [It still is not explicit what the author is
talking about. This assumption that the reader shares the author's

knowledge builds a potential bond – 'we're in this together!'] *Who'd have thought a pariah nation,* [the author assumes that the reader knows what a nation is. And the use of the world 'pariah' is a value judgment that may not have been shared by all readers, but since the author is assuming a bond between himself and the reader, the reader will likely agree with this judgment about a sovereign state that in fact has relationships with most nations of the world] *run by an authoritarian* [This is a reference to the president of Russia, Vladimir Putin, who enjoys an 80% approval rating among the Russian population – 'authoritarian' is a value judgement that is not shared by all] *who makes his political opponents disappear,* [Putin is a murderer] *could so easily hijack a great democracy? It didn't take much. A talented nerd can bring down a minnow of a nation. But this level of political crime requires more refined mechanics – you need everyone to play their assigned roles. You start with a stooge, a fugitive holed up in London, releasing stolen emails on the eve of the Democratic National Convention, in the name of 'transparency'. Cyberburglars rely on a partner in crime to pick up stolen goods. And WikiLeaks has always been there for Russia, a nation with no trans-parency.* [This entire passage is full of value judgements that are not universally shared but which the author believes are shared by the readers of the *New York Times*] …

The point here is not that newspaper pieces have deep and profound shared cultural knowledge (though what is 'deep' and 'profound' can vary from culture to culture and reader to reader). The point, rather, is simply that the author and the reader, for any of this opinion piece to work, must share cultural knowledge and, preferably, also a similar set of values. Or, if they do not share the values, both are adept at interpret-ing the values 'hidden' among the words.

Any article or opinion piece is, like this piece, saturated with unspo-ken judgements, opinions, values and knowledge that are never stated. The 'underdeterminacy' of this kind of information, its implicit nature, brings us full circle to the Banawá conversation this book began with. In human interaction the unstated is always crucial. Without culture, there is no language.

The British philosopher Paul Grice developed some helpful concepts for understanding the cultural and communicational presuppositions that underlie all human communication, which he referred to in the aggregate as the 'cooperative principle'. As Grice said, summing up his ideas, 'Make your contribution such as it is required, at the stage at which it occurs, by the accepted purpose or direction of the talk exchange in which you are engaged.' Grice makes this look like advice or a command, but it is actually intended as just a description of the cultural conventions underlying communication. We don't have to be taught these things. This is how we behave.

More accurately, Grice's principle of cooperation in communication is how we operate if we actually want the person(s) to whom we are talking to understand us. Every adept speaker follows the cooperative principle, as does every adept hearer. Their assumptions are further built on their unspoken cultural knowledge. Grice divided his cooperative principle into several 'maxims' which, when observed and, perhaps especially, when flouted allow for what philosophers and linguists refer to as 'conversational implicatures', things unsaid that are crucial for the meaning of what we have heard or spoken. The four maxims of Grice's principle caught on quickly among linguists, philosophers, psychologists and social scientists. These maxims are: the maxim of quality, the maxim of quantity, the maxim of relevance and the maxim of manner. And they are the perfect kind of discovery – simple and intuitively right.

The maxim of quality assumes that everyone will speak the truth. It presumes that neither the hearer nor the speaker will believe that anything will be presented as true if it is known to be false. It also assumes that no one will say that something is true if they lack adequate evidence. There are, of course, lots of untrue things said and many spurious postings of made-up facts on the internet. So Grice is not saying that people cannot lie. He is saying that hearers assume that they are not being lied to in the normal course of interaction.

In fact, in many languages, such as Pirahã, the actual verb in the sentence has to have a suffix that tells the hearer how good the evidence is for what the speaker is saying – inference, hearsay, or direct observation. So if one Pirahã asks another, 'Did so-and-so leave the village?' one possible response is, 'Yes, he did,' where the verb *did* would have a suffix appended that might indicate 'I saw him leave' or 'Someone told me he left' or 'His canoe is not here, so I infer that he has left' and so on.

To lie in any language, therefore, is to disregard the maxim of quality. Of course, since everyone lies, we know that there are times when we intentionally flout the cooperative principle. But even though we know that others flout the maxim and even though we know that we ourselves also flout it, if you tell someone something, they will initially believe it, other things being equal. In fact, English, like Pirahã, has verb forms that indicate the degree of truth or certainty in the things that we say. In English we call these markers 'moods'. So there is the indicative mood: 'John went to town'. There is the subjunctive mood: 'If John *were* to go to town'. There is the conditional mood: 'I wish that you would leave'. Or the imperative mood: 'John! DO IT!' All of these in their own way express the relationship of the word meanings to the truth of these meanings applied to the world around them. So indicative means that the world is the way it is being described. The subjunctive means that the speaker imagines that the world could possibly at sometime be the way it is being described. The conditional mood means that one would like to see the world in a particular way or does not want to see the world in particular way. The imperative means that one wants the hearer to make the world a certain way that it currently is not.

The next of the four maxims Grice lays out is the maxim of quantity. This is again in two parts. First, don't give any more information than the exchange requires. Second, relay all the information necessary for the current interaction. Let's say that someone asks, passing in the hall, 'Hey, how ya doin?' And you answer, 'At 8.30 I have a dental appointment. I have an irritable bowel issue today. I spent last night worried about my finances. Other than that, OK.' Or someone says, 'How did you meet your spouse?' And you answer with every detail you can remember about the concert at which you met. In both cases, English has a phrase to describe your answer: TMI – too much information. These answers exceed the information requested! This occurs when a speaker confuses irrelevant with relevant information. TMI violates the maxim of quantity. But there is another way to violate the maxim – too little information. Imagine that someone asks, 'What do you want to do tonight?' And imagine that the response is, 'Whatever.' Well, although that is not an unheard of answer to such a question, it is unhelpful. Such a vague answer fails to provide the amount of information expected in the exchange. Giving too much information or too little information deliberately are examples of flouting the maxim of relation (or relevance).

One of the more famous examples of defying a maxim is the letter of recommendation. Someone writes one but manages to say very little about the candidate's qualifications. They might offer the judgement that 'John has excellent handwriting'. Everyone knows in this case that the writer is flouting the maxim of quantity and implying that John is not qualified. What happened here? How does this implication emerge from the literal meaning of the words in the answer?

Or consider a spouse's violation of the maxim of relevance:

Husband: 'How much longer will you be?'

Wife: 'Mix yourself a drink.'

To interpret his wife's answer, the husband assumes first that she is following the maxim of relevance. Her answer, though it seems irrelevant, must be relevant. She is flouting the maxim (and maybe her husband's expectations as well). To understand how this otherwise irrelevant comment could be relevant, the husband must go through a set of cultural and personally based inferences. The husband concludes that his wife heard him and understood his question and that her answer, while not literally a response to the question, indicates that he should relax, she'll be ready in plenty of time. He shouldn't worry or bother her further. And the wife, for her part, has to know that the husband will be able to draw these inferences, based on her own inferences of how he will interpret her. Both of these examples – handwriting and fix a drink – work because they flout the maxim of relevance, 'be relevant'. Implicatures, how people interpret the flouting of maxims, are cognitively complex. They draw on a store of background cultural knowledge. For this reason interpreting conversation in light of the cooperative principle is highly culture specific. The maxims themselves, on the other hand, are probably found in all languages. Grice's maxims do not supplant culture. They assume it.

Consider Grice's maxim of manner. The interlocutors assume that each intends to be clear in their speech. 'Being clear' in this sense has four subcomponents. First, avoid obscurity. People believe that a speaker is making an effort to avoid ambiguity, to be as brief as possible while respecting the maxim of quality and to be orderly in their remarks. Again, these are not rules of etiquette for speech. Grice's claim is that his maxims are assumed by everyone when they talk. If someone uses an obscure expression, therefore, when their hearer expected a clear expression, they must mean something non-literal – they must be flouting this maxim for a purpose. So one infers the speaker's meaning. And if they come from the same culture or know each other well, they will in all likelihood infer correctly. Not always, however. There are frequently bad inferences that lead to confusion or misunderstanding.

People also interpret others charitably or uncharitably. That is, we believe someone means something good when we interpret them charitably. This is a favourable bias towards them or the situation's likely meanings. If someone says, 'That's an ambitious statement,' and their hearer interprets them charitably the hearer will assume that what is meant is something like, 'You really know your stuff. You are going places!' But if one interprets the same remark uncharitably they will quite possibly take this to mean, 'You bit off more than you could chew and your statement failed.' People use these modes of interpretation frequently in politics. They tend to interpret their own candidate charitably and their opponent's uncharitably. These types of interpretation are found all around – in marriage, sibling relationships, work and so on. The way people interpret what someone is saying is based to a large degree on the kind of relationship they have. A standard joke heard among university administrators is, 'Gee, if I say "good morning" to so-and-so, they will ask themselves "I wonder what he meant by that?"' If an employee either does not trust or fears his supervisor, this will colour his interpretation of what the supervisor says, however innocuous the supervisor's intent. If one believes in someone, trusts them and values their friendship, then if that person says, 'I will find you, no matter where you are,' the hearer will at least believe that the speaker will try hard to find them. If someone says, 'When I am elected president, I will make America more secure,' one is less likely to believe them. This is in part because they aren't known personally, or because no one believes any politician about anything. At least, they will be less credible than a 'normal person' would be, no matter what they are talking about.

Likewise, one's cultural experiences (however valid intellectually) can affect their interpretation of groups as well as individuals. If someone believes that all rich people are corrupt, then they will be less likely to believe someone rich when they say that the ability to make lots of money is good for the entire community, even if only one person does make a lot of money. If one believes that anyone who receives government assistance, welfare of any kind, is lazy or irresponsible, then if such a person says, 'I have to lie down,' one may be more inclined to see that as laziness than illness or being legitimately tired from hard work, even if one otherwise knows nothing about the person speaking.

This is all crucially relevant to language evolution. Even if *erectus* only said things like 'Eat. Drink. Man. Woman?' another *erectus* would have had to know what woman or group of women the speaker had in mind, when he might want to eat, whether he was telling him to get out of the way of his plans, among much other assumed information. Language is underspecified for meaning. Without culture, whether for *sapiens* or *erectus*, there is no communication. When he proposed the cooperative principle Grice revealed something about language evolution therefore that even he was in all probability unaware of. Only creatures that follow it can have language. There is no need to defend or criticise how we interpret others by a principle of charity. It is just a crucial characteristic of psychology in many cultures.

The relevance of all the above to language evolution is that even *Homo erectus*, *Homo neanderthalensis*, Denisovans and *Homo sapiens* would have – in the gradual construction of relationships, roles and shared knowledge bases – interpreted what people said, from the very first syllable uttered or gesture made, based on their view of the person and their understanding of their context. They would have 'filled in the blanks' of speech just as *sapiens* do. This is all a part of language that many linguists call *pragmatics* – the cultural constraints on how language is used. And these constraints guide our interpretations of others. They help us, as they helped other *Homo* species, to resolve the underdeterminacy of speech.

Another example of language being just good enough, depending on cultural knowledge for its use, is found in 'speech acts', the use of language to accomplish specific kinds of cultural goals. Oxford don John Austin and his student, Berkeley philosophy professor John Searle, introduced the analysis and terminology of speech acts into discourse about human language. Whenever anyone speaks to someone else they are engaged in an action of a very particular type. In fact, they are simultaneously occupied in many distinct acts. Austin talked about locutionary acts (what was said), illocutionary acts (what was meant) and perlocutionary acts (what happened as a result of what was said and what was meant). Each of these is important to the understanding of the nature and use of language and is therefore important to the understanding of the origins of language. And each of these must have been a feature of language since its beginning.

The locutionary act is speaking itself. If one asks, 'Where is Bill?' the very moving of the mouth, emission of air from the lungs and the arrangement and selection of the words used are all part of the locutionary act. But anyone performing this locutionary act is simultaneously performing an illocutionary act. An illocutionary act is the effect one intends their utterance to have. If one promises something they want their hearer to recognise that their promise is a promise. That is the effect one intends for their words to have. The illocutionary acts a person's words can accomplish include statements, commands, questions, or performative acts. The latter occurs when a minister, legally authorised to perform marriages, concludes a legitimate (non-faked, non-Hollywood) marriage ceremony with the words 'I now pronounce you man/husband/wife and wife/husband'. Many cultures make it such that uttering those words in the right context with the right authority (what philosophers refer to as the 'conditions of satisfaction' of the act) legitimates the couple's marriage. That is a performative act. These acts require more specific cultural supports than either statements or questions do. Therefore, they in all likelihood would have appeared later in the evolutionary record.

As languages have evolved, they have come to possess various types of illocutionary acts. One type is called representatives – acts that commit the speaker to the truth of the content of what she is reading or saying, such as a witness taking an oath in court or all graduating seniors reading the college pledge together. Another is referred to as directives – acts to get the hearer to do something. Directives include exhortations, direct orders, advice and requests. Then there are commissives, which are acts of commitment by the speaker, including promises and oaths of office. Expressives communicate attitudes and emotions, such as congratulations or apologies. Performatives are acts that by their mere performance bring something about, such as a

judge passing sentence. The list of speech acts recognised by researchers varies depending upon the author. But what is important about
them for language evolution is that they show how the use of language
is anchored in culture. *Homo erectus* is likely to have used representatives, directives, questions and commissives.

Finally, there is the perlocutionary act – what happens, or what one
hopes will happen, in the mind of my hearer when they speak. At the
end of an attempt to persuade someone, the effect of that person actually being persuaded or unpersuaded is a perlocutionary act. Thus a
translator of a book might say that a good translation should produce
the same perlocutionary effect in the readers of the translation that was
produced by the original. In other words, we communicate for effects,
perlocutionary acts. If one speaks or translates or otherwise engages in
the communicative enterprise they are hoping that their locutionary
act will be the right choice for the right illocutionary act to produce the
desired perlocutionary effect or act.

Notice that there is no way to have a language without perlocutionary, illocutionary and locutionary acts. If *Homo erectus* talked, then
they performed these acts. They would also eventually have learned to
be polite.

A *neanderthalensis* could have just grunted and demanded a piece
of meat. And that might even be what most of them did, because they
might not have had request forms as well developed as our own, at least
not initially. Politeness is interpreted as indirectness and gentleness in
the manner of letting people know what it is that one wants them to do.
It can also be used to report on the condition of one's body (such as, 'It's
been a long time since we've eaten,' or, 'Where's the bathroom?' both of
which indirectly report on a need), and seems to be sufficiently subtle
and nuanced that it would have come much later in the development
of language. As people began to learn that the use of force was usually
inefficient when interacting with those in their group they would have
begun to rely instead on the use of persuasion to get what they wanted.
The rise of politeness, adroitness and foresight in getting the other to
want to help us, or at least not feel forced to help us, led to the evolution
of indirect speech acts. These can take the form of speech or gesture
or body language, but their function is the same – getting another to
do something without actually saying what one wants them to do. This

is another example of the central truth about language, already stated earlier, namely that we do not say what we mean and we often do not mean what we say.

For instance, a man might prefer the air-conditioning thermostat to be set just about cold enough to hang raw meat, but his wife likes the temperature considerably warmer. She, being many times more polite than he is, let's suppose, might ask, 'Aren't you cold?' Or, 'What temperature do you have the thermostat set at?' Or even, 'I am cold.'

The man is supposed to infer, obviously, from the mere raising of the topic of temperature that someone, his wife, is dissatisfied and that he is being asked to do something about it. There may be body language accompanying such indirect requests that make the purpose behind the words even clearer. She may wrap herself in a blanket ostentatiously. But the main principle is what has been seen over and over – people never say all they mean. Hearers must use both knowledge about language and cultural knowledge (about thermostats in this case) to respond. If the man ask his wife, 'Do you want me to turn the thermostat up?' she might very well reply, 'No, I am OK.' But woe unto him if he does not figure out quickly that what she actually means is, 'Of course. Do I have to draw you a picture? Get a move on.' Of course, her indirect way of getting her request across is much more effective than a direct command.

I overheard a well-known philosopher and logician at a major university explain that if one was really angry with someone they could punch their lights out without fear of legal consequences. He explained further, only half-humorously, that if there are witnesses and one wants to get away with giving someone a drubbing, they should say exactly the opposite of what they're doing. Push someone over and say, 'Oh, you fell! Let me help you up.' Then one might kick them in the head while saying, 'Oh no, I tripped.' Then one might give them an elbow to the nose while uttering, 'Watch out! You're falling again. I have you! Don't worry!' The courts might see through this, but there is a chance that when the witnesses report what they saw the defendant's attorney will ask them what they heard. And then the prosecution case might fall apart. Why? The answer is not hard. Court systems often assume an inadequate theory of language, a theory that ignores what is *not* said and focuses only on what is actually uttered. But often what is not said is still communicated forcefully and what is said is just a decoy for something else.

The lesson, perhaps with some overkill, is again that language engages the whole person and the whole culture. In fact, it is more serious than this. One can make a case that no one can fully understand what anyone else says. We understand just well enough to get by. Or, as Herbert Simon said, language is just good enough – it 'satisfices' our requirements, but it is by no means the perfect system of communication. Yet when working hand in glove with culture, language is incredibly complex and rich. And such complexity and depth could only result from the evolution of the body and brain in tandem with psychology, language and culture for hundreds of thousands of years.

Conversational implicatures do not exhaust the contributions of the context to the interpretation of utterances, implicit information and speech acts, however. But they allow grammar to specify less information than required, leaving it to speakers to infer the remainder of the meanings from the context and culture of their exchanges. Grammar and culture work together in modern languages and this conjunction was almost certainly vital for the development of earlier human languages into modern languages. The grammar helps the inferences and interpretations of speakers, it does not determine them.

As linguists, philosophers and psychologists thought more about Grice's work on the cooperative principle, they discovered additional ways in which cooperation helps to structure language. In the mid-1990s Dan Sperber, a cognitive scientist at the CNRS (Centre national de recherche scientifique) in France, and Deirdre Wilson, Professor of Linguistics at University College London, collaborated to produce the theory of human interaction known as 'relevance theory'. Relevance theory explores applications of the cooperative principle beyond conversation. Relevance theory, like Grice's work, sees language forms and interactions as governed by a culturally based pragmatics.

All of the work in pragmatics (which is partially the study of how the context in which one is conversing determines which interpretations are appropriate), sociolinguistics (how language and society each affect the other) and additional disciplines of scientific research that look at language in use, rejects the so-called 'conduit metaphor' of language. This represented a tremendous advance in the formalisation of

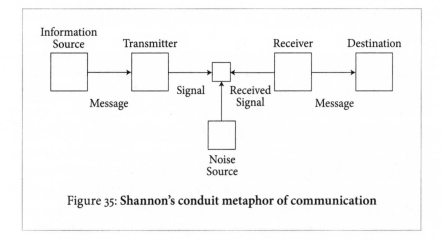

Figure 35: **Shannon's conduit metaphor of communication**

the theory of communication. And the highly respected name usually associated with its formulation is mathematician Claude Shannon. This metaphor is the idea that communication is linear, where a thought arises in the speaker's mind. Next, the speaker selects a grammatical form for that idea. The speaker then transmits the idea-in-the-form to the hearer's ears (or eyes, if using sign language). The hearer takes apart the grammatical form and is left with the meaning that the speaker intended. This is represented in Figure 35.

In the late forties and early fifties, when Shannon was writing his most pioneering work, there was almost no research on the formalisation of communication theory (though certainly work such as that of Alan Turing on the code of Nazi Germany's Enigma machine was foundational to thinking in the area). As a researcher at Bell Labs it was Shannon's job to work out an understanding of communication that could be translated into mathematical models in order to help Bell make telephones ever more effective and efficient.

As Figure 35 shows, Shannon's conceptualisation of the communication problem leaves no space for external influences on the process of communication, such as culture or context or gestures or intonation. It is almost as though all one needed to communicate was two brains, two vocal apparatuses and two auditory systems. Shannon, who knew Alan Turing and other founders of computational theory, developed his system in 1948 and it has been accepted as foundational to most other work in the fields of cognitive science, electrical engineering, linguistics, psychology, mathematics and other fields since then.

But Sperber and Wilson, following on the work of Grice, Searle, Austin and many others, said in effect, 'The conduit metaphor has had its day, but it really doesn't capture what happens in human communication at all. The conduit is just a set of points in a much larger set of events and processes that underlie communication.' In the view of relevance theory, any time a story is told or a conversation is had or a sentence is uttered, there is always a context of utterance. Not only this, the set of interlocutors assumes that each member of a particular communication event – a storytelling, a speech, a conversation – knows the context as well as the relevance of the context to what is being said and is therefore able to comprehend and respond appropriately, in a relevant manner. One doesn't say something unless it is relevant. One doesn't imply or interpret something unless it is relevant to the context in which the discussion is taking place. Thus if someone says something to someone else, the hearer will assume that it is relevant and they will therefore make an effort to understand what was said or written.

As a concrete example, what is the relevance of a discussion of relevance theory or Grice's cooperative principle in a book about language evolution? The reader makes the effort to process the sentences that they are reading but does so assuming that these sentences will be connected in some way to our general topic of discussion – language evolution. And indeed they are. But before turning to the evolutionary significance of these ideas, one more item on the topic of 'language emerges from context' should be discussed – conversation. The apex of our evolution as a species.

Conversation seems innocuous enough. I say something. You say something. We think about what the other has said, or not. Then we finish and say goodbye. Something like that. This simplistic view of conversation is not entirely wrong. It is just fundamentally incomplete. Therefore, to understand how language evolved to be 'just good enough' we need to look at conversation. Here is a snippet of conversation from a couple I observe regularly:

WIFE: What time are we going to pick up Miguel on Thursday?
HUSBAND: I already told you.

WIFE: Yes, but that is when you were planning to pick him up after work. Now we are just driving directly to pick him up.

HUSBAND: The time doesn't change as a result.

WIFE: But just give me a time that we will need to leave, so I can call the dogsitter.

HUSBAND: Whenever you want to leave in order to get to his place between ten and twelve.

WIFE: Why can't you just tell me the time? Why are you playing these games?

HUSBAND: I gave you a range and from that there should be no problem inferring the time.

WIFE: I don't even care if we go.

HUSBAND: Fine.

The interpretation of this depends on the literal meanings of the words, but also on the personalities involved, the cultural notions of time, the complication of the drive to pick up the person in question and the fact that both people are tired. One wants things stated clearly and precisely. The other likes to be less committed to times (in some situations – in other situations their roles are in all likelihood reversed with regard to time and precision). Moreover, when the wife says, 'I don't even care if we go,' that is not intended literally. Both interlocutors are keen to spend time at the beach with the friend being discussed. This line is uttered to make a point, namely, 'If you cannot simply answer my question with a specific time, then my feelings must not be that important to you.' The final, petulant response, 'Fine,' is also not literal in this case. It is just to say, 'If you are going to make a big deal of this, I am not going to respond as expected.' Aside from their amusement value, such exchanges show how the interpretation, the utterances and the exchange as a whole can only be understood through extensive external knowledge of culture, local circumstances, the relationship between the interlocutors and their individual personalities. Shannon's conduit metaphor is of very little help here.

Language, psychology and culture, again, co-evolved to produce the contextual linkage between the world, personalities, cultural understandings, current events and so on that make full interpretation of language possible. Moreover, we see a great deal of variation in how this is done across different languages. Consider a few exchanges and then a discourse on how to make arrows in Pirahã

Greeting in Pirahã:

Ti soxóá.
'I already.'
Xigíai. Soxóá.
'OK. Already.'

Or this exchange:

Ti gí poogáíhiai baagábogi.
'I give you a banana.'
Xigíai.
'OK.'

Or another, this one to communicate that you are leaving, where English speakers might say something like, 'I am leaving now, goodbye.'

Ti soxóá.
'I already.'
Gíxai soxóá.
'You already.'
Soxóá.
'Already.'

Pirahã doesn't have words for thanks, goodbye, hello and so on, what linguists refer to as 'phatic' language. They allow the context to determine the bulk of the meaning for obvious things like leave-taking and arrivals. They see no need to say thank you, in part because every gift carries an expectation of repayment. If I give you a banana today, you should give me something, such as a piece of fish, when you are able. This is not stated. It is culturally presupposed. So 'giving' is actually a form of exchange by and large. If a gift in the English sense is intended, that is, without expectation of repayment, I would say:

Ti	gí	hoagá		poogáíhiaí	baagáboí.
I	you	contraexpectation		banana	give (literal translation)

'I (against your expectations) give you a banana.'

The addition of the word **hoagá**, 'contraexpectation', renders the meaning of the entire sentence in effect: 'Contrary to what we normally do, I am just giving you this.' This may sound a bit circumlocutious. But it results from the Pirahãs' assumptions being quite different from those of most Western cultures. In any case, there was no need in the course of the historical development of the Pirahã language to develop vocabulary for such purposes for exchanges, only for giving without expectation of repayment.

When *Homo erectus* began to communicate orally, it would have been even less necessary to develop a language that expressed everything. After all, language would have emerged from a society that was increasing in complexity and communicating with body language, perhaps vocal interjections, hand signals, maybe some representative spatial drawing with a stick in the dirt as they planned hunts. But let's assume that a local *erectus* community somewhere had settled on the rudiments of some symbols. One cannot expect the inventors of a single symbol to try to pack all the meaning in the world into that symbol. In fact, one could not expect that even for a group of symbols, no matter how many symbols in that set. There is too much information in the background context, too much in our memories that we use to interpret but don't actually say (and often don't even know we know, or don't know that we are using) for mere symbols, however enriched by gestures and intonation and body language, to say everything. Thus it is clear that as language evolved, speech acts, indirect speech acts, conversations and stories depended heavily on cooperation, implicit (unspoken) information, culture and context. It is the only way that language has ever worked.

Part Four

Cultural Evolution of Language

12

Communities and Communication

In any situation when two people are talking, they create a cultural structure. Our task, as anthropologists, will be to determine what are the potential contents of the culture that results from these interpersonal relations in these situations.

Edward Sapir

IN THE COLONIALIST EXPANSIONS and explorations of European nations, communities very different from Europe were discovered. These newly discovered communities shocked ethnocentric Europeans. Their radically different appearances and ways of life raised the issue of whether all creatures that look like humans are indeed fully human. Do they all have souls? Many Europeans thought not. At least they believed in the inferiority of these people they had just 'discovered', leading to their justification for exploitation, colonialism and enslavement of them. Did they all come from the same source as us, God? Are some varieties of humans superior to others? Anthropological and comparative linguistic studies arose from such questions, questions fundamental to the understanding of language evolution culturally and biologically. And these issues continue to be questioned by some.

One prominent example of European thought about cultural and linguistic differences was Sir William Jones, who served the British Raj as legal counsel in the late eighteenth century. Jones was more than a common solicitor, however. For one thing, he was a man of radical politics, strongly supporting the efforts of his friend and one-time co-author Benjamin Franklin for American independence. Jones also studied the social systems of India. But most importantly for intellectual history, Jones was a linguistic prodigy, speaking thirteen languages fluently and another twenty-eight reasonably well, as the story goes. He put this linguistic brilliance to use not only by talking different languages, however. He also wanted to understand these languages scientifically. Most importantly for our story here, Jones also searched for evidence about the historical connections between these languages.

During Jones's survey of data from various sources, he experienced one of history's most important 'Eureka' moments. He had rediscovered a fact first noticed more than one hundred years earlier by German Andreas Jaeger in 1686 and again in 1767 by the French Jesuit missionary to India, Gaston-Laurent Coeurdoux. Though the work of Jaeger and Coeurdoux was largely ignored, Jones's independent observation of the same fact was to reverberate through the centuries as one of the most important findings in the study of human communication. This insight was that Sanskrit, Greek, Latin, Gothic (German-related languages) and Celtic all traced their origin back to a common ancestor. They were sister languages. Their mother language – also the mother of many other sibling languages awaiting their turn to be discovered or otherwise enter the family tree – came to be known as *Proto-Indo-European*. With Jones, Jaeger and Coeurdoux the study of language origins began in earnest.

Then almost one hundred years later, near Weimar, Germany, another important tool was developed for researching language origins. In 1850 twenty-nine-year-old German philologist August Schleicher published a book in which he claimed that human languages should be studied as organisms, on a par with biological organisms, related to one another by genus, species and variety – the same sorts of relationships we now understand to hold between all flora and fauna. Schleicher made the case that the best way to represent the evolutionary relationships between languages was by 'tree diagrams'. With this proposal he not only made an enormous contribution to the history and evolution of languages but also introduced the concept of 'natural descent' – nine years before Darwin published his *Origin of Species*.

Schleicher's and Jones's work inspired others to think deeply about the relationships between languages. It became apparent that, using the method of constructing language trees that began to be developed in India, Germany, France, England and elsewhere, we could look back into time in our search for where and when specific languages originated. It was eventually discovered that Indo-European was the mother of most European languages. And it was then discovered that this was also the mother of non-European languages such as Farsi, Hindi and many others. The question thus naturally arose whether we could discover the mother of Indo-European itself. We know now that Indo-European began to split into the modern European languages approximately 6,000 years ago. Can we go further? Ten millennia? A hundred? Could we actually use the methods of comparative and historical linguistics to reconstruct the first language ever spoken?

Most contemporary linguists respond to this question with a firm 'No'. The methods Jones's work called attention to seem to hit the wall at about 6,000 years. To go deeper we will need the methods of other fields, such as palaeontology, archaeology and biology – and we'd need what we probably can never have, preserved samples of languages.

But the question remains. If we were able to travel back further than those 6,000 years, where would we end up? Would the quest of Jones, Schleicher and others take us back to a single language at the root of one enormous tree of human languages? Some people think so. The late Stanford professor Joseph Greenberg claimed that we could trace all human languages back to a single source, which he and his followers labelled proto-*sapiens*. But other scholars say no. They maintain that there are many trees all going back to different prehistoric communities of hominids. Greenberg and his disciples believe in monogenesis, the hypothesis that there is only one beginning – one mother tongue – for all human languages. Others advocate polygenesis, that there are multiple evolutionary beginnings for modern human languages. These people argue that the ancestors of modern humans left Africa speaking different languages. Different communities of speakers developed different languages that in turn are the sources of all modern languages. Choosing the best hypothesis, monogenesis vs polygenesis, is just one of myriad problems we face in trying to reconstruct the evolution of human languages.

We know that other methods, other sciences beyond linguistics, can take us further back in time. But can they transport us to the beginning of human language? Can we know anything about who told the first story? Or who first said, 'I love you'? The romance and the science come together in this story of the search for human language origins.

It is a tale replete with scientific controversy and marked by frustrat-
ingly slow progress towards the ultimate goal of knowing how humans,
but no other species, moved from mere communication to language.
Although historical linguists believe that the methodology of their
field is unable to reveal much once we go beyond 6,000 years ago, its
major insight, that languages change over time due to a combination
of cultural and linguistic reasons, is essential to an understanding of
language evolution.

Historical (or 'diachronic') linguistics, the field virtually launched
by Jones's work, is the field dedicated to understanding how languages
change over time. For example, English and German were once the same
language ('Proto-Germanic'), as were Spanish, Portuguese, Romanian
and French (Latin). And we know that Latin and Proto-Germanic were
themselves one language some 6,000 years ago, Indo-European. The
science of how languages drift apart as these languages have is one of
the oldest branches of the study of language and it is relevant to the field
of language evolution. After all, if *Homo erectus* underwent evolution
to become *Homo sapiens*, maybe the language of *Homo erectus* also
changed into the languages presently spoken by *Homo sapiens*. Any
change in *erectus* languages, however, would be beyond the science of
historical linguistics. That is because *erectus* lived much longer than
6,000 years ago. Even one of the main tools used by historical linguists
for dating when one language in all likelihood split to become another,
glottochronology (literally, 'tongue time'; referred to by some as 'lexi-
costatistics') is of no assistance to here. Glottochronology, invented by
linguist Morris Swadesh, assumed that there were some vocabulary
items (such as parts of the body, words for sun, moon and others) that
were less likely to be borrowed. He therefore came up with a list of two
hundred words or 'lexical items' which he considered represented the
words least likely to change. A mathematical formula was proposed
and developed, based on the rate of change of the words in his list, to
predict the rate at which these most resistant-to-change words might
in fact change over time. The formula was tested and deemed to have
87 per cent accuracy in known cases, such as the Indo-European languages. Though many linguists still are highly sceptical of the method,
it does seem to be useful. But it is not able take us back further than the
6,000 years wall. So it is not a tool for language evolution.

What it and the entire field of historical linguistics do show, however, is that languages continue to change. In fact, linguists recognise that change in modern languages is largely the result of a form of linguistic natural selection that would have certainly been operative in the first languages ever spoken. All languages change all the time. They change because of geographical separation (think 'genetic drift'), or from differing preferences of age, economy, race and many other factors. And these forces in one form or another mean that the languages of *Homo erectus* began to change as new communities were formed. Much of historical linguistics boils down to the idea 'You talk like who you talk with.' Once you stop talking to a group of people, you will eventually stop talking like them. Or at least your group will. This is why each time we reach a major river or mountain range in Europe, we are likely to find distinct languages on each side that were once the same language. As for English and German, English was born from German after the crossing of the English Channel by the Saxons.

Because language is a cultural artefact, we must understand what culture is in order to understand language. So what is a culture, then? Is a culture like a football team? Or perhaps like an orchestra? Or is culture simply the overlap in values, roles and knowledge of individuals who live together and talk together? The larger issue is how cultures hang together at all. In other words, in what sense might the American motto *E pluribus unam** describe American culture? Since I claim that culture is an abstraction, it can only be found in the individual. It is the result of a 'gestalt'. From its various individual members, a culture emerges which is greater than the sum of its parts. Understanding culture has profound implications for understanding language evolution.

I have in my own work developed a theory of culture in which the individual is the bearer of culture and the repository of knowledge, rather than the society as a whole. I want to look at the effects of culture on the nature of national and local societies and individuals and their languages, via examples such as the role of the teacher in the classroom, the organisation of businesses and the organisation of societies. The three ideas of my work that are most important for language evolution are values, knowledge and social roles.

---

*Out of many, one.

As a further illustration of the importance of culture in language, consider the following interactions between two linguists:

A. 'Colourless green ideas sleep furiously.'

B. 'They sure do.'

The general population of English speakers may have no idea what A's utterance means. But if A and B are members of the culture of linguists, then they know that this is a famous example sentence in Chomsky's early writings that is designed to show that a sentence can be grammatical yet meaningless. For the two linguists, A's sentence is an insider joke and B's response a humorous rejoinder. The function of the exchange might be largely phatic, simply to say, 'Hey, we are both linguists.' What is often overlooked, though, is that B's reply shows that 'Colourless green ideas sleep furiously' is not, in fact, meaningless. It tells us that, whatever green ideas are, they sure do sleep furiously. In other words, because of the cooperative principle, all people will believe that an utterance has meaning and will work to attribute meaning to it, regardless of what words it is composed of.

Now consider the following. Persons C and D are watching the New England Patriots play the Miami Dolphins. The Patriots take the lead. C and D both yell, 'Yes!' and 'high five' one another. In this joint action they show knowledge that there is a game of football, knowledge of how this game is scored, shared value-ranking for the Patriots relative to the Dolphins, knowledge of what 'high fives' are like and what they are for, knowledge that they are both 'rooting' for the same team and reinforcement of all of the above.

From such culture activities come knowledge, community belonging and shared communication. These exemplify the role of living in a culture and speaking a language in constructing our identities and our societies. From these actions the individual assembles his or her own experiences and an ability to understand the actions and speech of their fellow culture members. As an example of how unspoken much of culture is, it is worth reviewing one of the many failed treaty attempts between North American indigenous peoples and their European-immigrant conquerors.

A historical incident, the famous Treaty of Medicine Lodge, signed in 1867 between the Arapaho, Kiowa and Comanche peoples and the US government at the Medicine Lodge Creek in Kansas, was simply one of many failed communications between European and indigenous communities based on misunderstandings caused by the tacit cultural information that is required to interpret what language leaves unsaid. A serious and war-threatening misunderstanding grew out of two distinct cultural interpretations of this deceptively simple-looking treaty. The indigenes expected one thing. The government expected another. And both were right according to the language. This is a common source of misunderstanding between individuals, across cultures and internationally. Such misunderstandings boil down to culture's role in filling in the gaps that language is missing and underlying the interpretation of language itself. I want to look at the Medicine Lodge misunderstanding in more detail.

For more than a century, anthropologists have bickered about the definition of culture. The members of a given family, community, society, or nation clearly share some knowledge, some values and some relationships. They may talk alike. They may dress alike. They may show disgust at similar things. They may all drink coffee from their saucers.

So the question arises naturally, 'What is culture?' Culture is the tacit knowledge and overt practice of social roles, values and ways of being shared by a community. Each of us has many different roles. I am a father, a teacher, an administrator, a husband, a shopper, a patient and a researcher. Each one of these roles is recognised by most members of my community. To the degree that they are recognised, my community shares knowledge of these components of culture with me. Culture distinguishes and shapes us, even when our roles may seem universal, such as 'father', and so might appear independent of culture on the surface. But although there are Italian fathers and American fathers, the concept of 'father' is not identical in every culture. It seems likely that between any two cultures, fathers will have overlapping but never identical roles. Even fathers of ostensibly the same cultures vary in the nature of their roles at different times.

Some societies may believe that fathers should support their families. In such a society, it may be assumed that fathers have a responsibility to provide food, clothing and shelter for their children. And, in Western societies at least, both the society and many fathers themselves believe that it is good for fathers to help their children with schoolwork, heavy lifting and tasks in general too difficult for children to do alone. Fathers of other generations may share exactly these beliefs and values. But these values are not identical across different cultures. A Pirahã father will not often pick up a child that has injured itself to offer comfort. He will expect the child to work hard and not complain on long treks through the jungle and will not offer assistance in many cases that the American father would. And his individual values emerge partly from the values of other members of his society.

And, of course, across different generations, fathers may differ profoundly. Values shared by many of my father's generation included corporal punishment, the expectation that women did the bulk or all of the housework, the belief that their wishes and orders would be carried out without question and the attitude that their children were not deserving of respect or of a voice in family affairs. These fathers might regularly side with teachers against their own children in disputes. They considered the child and all its resources as mere extensions of themselves or their possessions. The fathers of the generation of my children, on the other hand, usually avoid corporal punishment, see their family as a unit of equals, know they ought not to believe that their desires should be the only or even the main ones heard, often help clean the house, would almost always take their children's side in a school dispute and so on. Being a father in the 1950s was considerably different from being a father in the twenty-first century. This is because the cultural role of 'father' is defined by shifting cultural values.

If my quick summary of the evolution of fathering over the past few years is on the right track, then the fact that the changes affect entire generations similarly indicates sharing of values – culture. This is part of what it means for a group to have a culture. All cultural roles show similar diachronic, geographic, economic and other shifts across time, space, or populations. If we move from roles to beliefs or from beliefs to shared concepts, to shared phenotypes (a phenotype is the visible appearance and behaviour of an individual), shared food and shared music, we will find many examples of shared knowledge producing distinct cultures.

In part these shared mental items emerge because over the course of one's life, each accumulates experiences, lessons and relationships. These are all in a sense assimilated into our bodies and minds. People who grow up in the same community have similar experiences – climate, television, food, laws and values (such as fat is bad, honesty is right, hard work is godly). Episodic and muscle memories hold our various experiences together as cultural experiences embed themselves within us. Arguably our 'self', or at least our 'sense of self', is no more than this accumulation of memories and apperceptions.

How does one recognise another as part of the same culture? Indexes are readable by members of any culture, in fact by members of most species. They are clues to the environment necessary for survival across many life forms. And thus it is known that the ability to connect a representation to a form is an ancient ability of the genus *Homo*. Humans were never without it. Icons, on the other hand, require more. Whether making an icon or simply collecting one, the reader of the icon must understand that it physically resembles what it represents. Understanding, whether indexes, icons, or symbols, is an intentional act, directly or indirectly. (This is because to understand requires at least tacit recognition of connections between the sign and the thing it refers to.) An index is itself, however, non-intentional. One doesn't plan for a footprint to be connected to man. It just is.

The ability to interpret cultural information comes slowly. Everyone is born outside of culture and language. We are all partially aliens as we emerge from our mothers' wombs ('partially' because learning about our new culture and language begins in the womb). When we are born, we are outside our mother's culture looking in. Our senses provide information. But it takes time to interpret what we are seeing, feeling, tasting, touching and hearing.

Examples of misunderstandings caused by culture are plentiful in the history of the Native American and 'manifest destiny' in the nineteenth century. Communication between them failed many times due to an inadequate appreciation for each other's culture, just as communication between governments in the twenty-first century often do.* An example alluded to earlier is the 1867 Treaty of Medicine Lodge.

This treaty was ineffective from the beginning. For once at least an official treaty with the Indians was invalidated not because of

---

*The idea that the expansion of the United States throughout the American continents was both justified and inevitable.

dishonesty on the part of the government but because the signatories failed to realise that language, whether spoken or written in treaties, is merely the visible portion of an invisible universe of understanding that derives from the values, knowledge and experiences – the cultures – of individual communities. Though people might read the same words in a treaty, as in all communication our interpretations are slaves to our assumptions, based on background beliefs and knowledge that the literal meaning of the words rarely conveys.

In this case, the treaty called for the government to provide food to the Indians so that they could feed their families through the winter months. The Indian agency was responsible for providing the food. Congress was responsible for ratifying the treaty that was signed. Each in turn depended on other cultural institutions, all with their own deadlines and priorities. The Indians gave no thought about ratification. But they should have, because when they arrived to collect their provisions, prior to ratification of the treaty, the pantry was bare. The government had provided nothing because the treaty had not yet been approved. Regardless of the reason, the Indians felt betrayed.

On the other side, the government thought that the Indians, when they had agreed to live in the reservations, would now consider themselves bound to stay there in perpetuity and to forever abide by the 'law'. Perpetual obligations to anyone other than their own families were foreign to the Indians' values and understanding of the way the world worked. They could never have legitimately made the commitment expected of them. It made no sense. The government could not have cared less about Indian interpretations rooted in their very different cultures. But it should have. Comanche chief Quanah Parker, present at this ill-fated gathering, at least learned from the experience. In his future dealings with whites he learned to respect the importance of the dark matter of the unsaid. He subsequently inquired about every potential assumption that he thought whites might be making before signing future treaties (though no one outside a culture can ask all the right questions).

Treaties often break down over cultural misunderstandings. But there are plenty of everyday examples of culturally induced language breakdowns as well. If you tell someone, 'We should do lunch sometime,' what do you mean? In your community this might literally mean that we should now, the two of us, plan a meeting for lunch at a restaurant. Or it could mean instead, 'I have to go now. I don't have time for this conversation.' The interpretations of the interlocutors will be based on their relationship, their knowledge of each other's cultural and personal expectations and by monitoring of the looks of anyone else standing around, as well as each other's expressions and gestures. The meaning of what is said is never based merely or even mainly on the words spoken in a conversation.

The point is that human language is not a computer code. Humans did not gain a grammar first and then figure out its meaning in a particular culture. Culture, grammar and meaning each imply the others in human language. Languages and psychology run their wells deep into cultures. No artefact in human languages or human societies can be understood except by means of the culture in which it is interpreted. Understanding the nature and role of culture in human behaviour, language and thinking is essential for comprehending the evolution of human language.

There are various arguments that modern researchers occasionally employ to deny the existence of culture and omit it from the construction of their theories of human thinking, behaviour and language. This may be because of their training or their adopting a poor definition of culture. Some theorists, both in linguistic theory and in language evolution, disregard more than a century of anthropological studies making a powerful case that culture is necessary to explain the human animal.

Each community of *erectus*, *sapiens* and *neanderthalensis* would have developed familiarity with one another, a sense of 'togetherness', leading to shared values, social roles and structured knowledge ('structured knowledge' means knowing not only lists, but also how things on the lists relate to one another). Sharing such things would have brought them to a degree of cultural homogeneity. Perhaps they had a common symbol that was used in times of difficulty, such as an emergency warning, either spoken or some sort of signal (smoke signals helped identify some Native American communities). Perhaps not. But each band of travellers necessarily shared a spirit and a culture that underwrote their communication.

Modern organisations work hard to develop slogans, chants (such as the 2016 Republican chant 'Lock her up!' directed against Democratic candidate Hillary Clinton), anthems, phrases for the population as a whole. When the group proclamation becomes individuals' value, the social and the individual become linked. This forms culture and alters language. Words take on new meanings or new words and new meanings are born. Culture changes bring language changes.

Culture, patterns of being – such as eating, sleeping, thinking and posture – have been *cultivated*. A Dutch individual will be unlike the Belgian, the British, the Japanese, or the Navajo, because of the way that their minds have been cultivated – because of the roles they play in a particular set of values and because of how they define, live out and prioritise these values, the roles of individuals in a society and the knowledge they have acquired.

It would be worth exploring further just how understanding language and culture together can enable us better to understand each. Such an understanding would also help to clarify how new languages or dialects or any other variants of speech come about. I think that this principle 'you talk like who you talk with' represents all human behaviour. We also eat like who we eat with, think like those we think with, etc. We take on a wide range of shared attributes – our associations shape how we live and behave and appear – our phenotype. Culture affects our gestures and our talk. It can even affect our bodies. Early American anthropologist Franz Boas studied in detail the relationship between environment, culture and bodily form. Boas made a solid case that human body types are highly plastic and change to adapt to local environmental forces, both ecological and cultural.

Less industrialised cultures show biology-culture connections. Among the Pirahã, facial features range impressionistically from slightly Negroid to East Asian, to Native American. Differences between villages or families may have a biological basis, originating in different tribes merging over the last 200 years. One sizeable group of Pirahãs (perhaps thirty to forty) – usually found occupying a single village – are descendants of the Torá, a Chapakuran-speaking group that emigrated to the Maici-Marmelos rivers as long as two centuries ago. Even today Brazilians refer to this group as Torá, though the Pirahãs refer to them as Pirahãs. They are culturally and linguistically fully integrated into the Pirahãs. Their facial features are somewhat different – broader noses, some with epicanthic folds, large foreheads

– giving an overall impression of similarity to East Asian features.* Yet body dimensions across all Pirahãs are constant. Men's waists are, or were when I worked with them, uniformly 27 inches (68 cm), their average height 5 feet 2 inches (157.5 cm) and their average weight 55 kilos (121 pounds). The Pirahã phenotypes are similar not because all Pirahãs necessarily share a single genotype, but because they share a culture, including values, knowledge of what to eat and values about how much to eat, when to eat and the like.

These examples show that even the body does not escape our earlier observation that studies of culture and human social behaviour can be summed up in the slogan that 'you talk like who you talk with' or 'grow like who you grow with'. And the same would have held for all our ancestors, even *erectus*.

People unconsciously adopt the pronunciation, grammatical patterns, lexicon and conversational styles of those they talk with the most. If one lives in Southern California, they might say, 'My car needs washing,' or, 'My car needs to be washed.' But in Pittsburgh, they are more likely to say, 'My car needs washed,' or, 'My car needs to be washed.' There is a grammatical contrast between the two dialects. The Southern Californian dialect requires the present participle form of the verb, whereas the Pittsburghese dialect requires the past-tense form of the participle. Both cultures converge in the 'to be' construction. As another example, if you talk to people of my generation you are likely to say, 'He bought it for you and me,' whereas if you talked mainly with members of a more recent generation, you might say (ungrammatically), 'He bought it for you and I.'

Although imitation is a major cultural force, always pressuring a society towards homogeneity, it is not the only force. There is also innovation, which pressures societies to change. Imitation, though, is the seed of culture. The structures and values constitutive of culture take time to evolve. These structures and values emerge partially through

---

*Obviously DNA studies would be interesting and necessary scientifically before saying anything with confidence on this score, but it is difficult politically to carry out such studies because in Brazil those delegated to protect indigenous peoples are wary of anything that could be perceived as racist studies, especially studies carried out by 'gringo' scientists.

conversational interactions, which include not only the content of speech, but also perspectives on right and wrong actions or thoughts, acceptable levels of novelty of information or form of presentation and levels and markers of conformity. This happens as *people talk like who they talk with*.

In other words, people who interact become more alike. Raise two children together and they will be more alike than had they been raised apart. They will share values that children raised apart do not share and they will, at least early on, share knowledge structures that are more similar than had they been raised apart. The more people talk together, the more they talk alike. The more they eat together, the more they eat the same foods in the same way – the more they eat alike. The more they think together, the more they think alike.

The more people's values, roles and knowledge structures overlap the more connections they share and, therefore, the stronger is their connection in a cultural network. Thus they can form a generational network, a CEO network, a rap-lovers network, a 'Western culture' network, a stone tool flakes network and an industrialised society network, or even a *Homo sapiens* network, so long as they share values, knowledge, or roles.

This is recognised by many people when they claim that 'people are all alike'. This is a common truism. Culture is only superficial, it is thought. We do all share some values. Likewise the other extreme, represented by cultural relativists, is also right when it claims that no two cultures are alike. No two cultures or individuals share all the same values, all the same social roles, or all the same knowledge structures.

What were the components that were changing early *Homo* from bands of individuals into cohesive cultures? First, there were values. These are the assignment of adjectives of morality for the most part (more clarification of values will come directly) to specific actions, entities, thoughts, tools, people and so on. They are also statements about how things should or should not be. To say, 'He is a good man,' expresses a value. This can be broken down into finer-grained values such as, 'He treats his children well,' or, 'He is kind to stray animals,' or, 'He gave me a ride home,' or, 'He is polite,' etc. Values are also seen in the tools we choose – a bat instead of a gun for home defence or a machete instead of a hoe for digging vegetables in the garden. They are seen in the use of our time. Value sets are vast and varied.

My definition of culture also includes the phrase 'hierarchical knowledge structures', which refers to the idea that human knowledge, at least – perhaps this also applies to other animals – is not an unordered set of ideas or skills. What we know is broken down in various ways according to context. All is structured in relation to all. And this hierarchy inescapably produces a gestalt output, meaning that the sum of what we know forms a system that is greater than merely all the things we know put together. Just as a symphony is greater than a mere list of all of its notes.

In my understanding of culture the idea of 'social roles' is useful to describe actions as conforming to a particular position one occupies in a culture. Any grouping of people will be defined by its values, the knowledge structures it devolves from and develops and the expected duties of each of its members by virtue of their membership classification.

To take an example from business managers in North America, China, or the United Kingdom, these folks will differ in many of their values, administrative knowledge and more, but in their social roles (independent of what they are called), they will necessarily share some aspects of administrative knowledge and values. In a sense, then, there is an international management culture, broken down into national and company-specific, local subcultures. Likewise, in higher education there is careful watch over expected cultural values in the form of different accrediting bodies. Accreditors allow schools to operate insofar as the schools share and implement the agencies' values.

As *Homo* species traversed the earth, they too shared values with all of those in their species. In fact, given the relative homogeneity in *erectus* lives – all were hunter-gatherers – the cultures of different *erectus* communities would have been, superficially at least, quite similar. Of course, there were also important differences. Some of these differences would have resulted from the different ecologies of separate *Homo erectus* bands. Some lived in cold climes, others in the tropics, while still others braved the sea to live on islands. These were the forces that led the original immigrants from Africa to the formation of distinct cultures.

Most studies of values fail to provide a theory of the relationships between values and because they do, they too often assume that all values are universal, though aside from biological values there is no evidence to support this.

The ranking or prioritisation of values is easy to illustrate. Suppose that we are comparing the values of the inhabitants of two cities, say Paris and Houston. Let us further assume that Parisians and Houstonians value 'good food', however they define 'good' and 'food' locally. And let us suppose that both of them value being in good shape. Now, for the sake of discussion let's assume the following rankings (the symbol '>>' means that the value on the left 'outranks' the value on the right):

Parisians: Good shape >> Good food
Houstonians: Good food >> Good shape

In this hypothetical scenario, it is more important to Parisians that they be in good shape than that they enjoy good food. Though they do enjoy good food, they will not overeat if that causes them to no longer be in good shape. Good food takes a back seat to health and the waistline. To the made-up Houstonians, however, being in shape is not as important as the enjoyment of good food. Abs and glutes are less important, say, than batter-fried okra and chicken. It seems fair to say that these value priorities would produce different body shapes, especially if we add to this ranking a finer analysis of what each group considers to be 'good food'. Houstonians might prefer fried chicken and mashed potatoes. The French might like instead coq au vin, etc. But it would be correct to say that the two cities have the same values. In this case it is not the values but their relative ranking that makes the difference. So we need some idea not only of what a group's values are, but also of the prioritisation of the values. One cannot say what a group's values are, though, without studying them carefully. So we cannot infer much about the culture of *Homo erectus* communities. But they would have had values and their values would have shaped their daily lives, and some of these would have been more important than others.

In the 1950s Kenneth Pike began work on a 'grammar of society'. He suggested that the principles of human grammars are the organising principles as well for 'grammars of culture'. In this sense, a culture is partly grammar-like. Like any grammar, a culture-grammar can only be proposed based on solid methodology and rigorous testing of hypotheses.

Society and culture are, of course, more than merely grammars – but they are connected and constructed in grammar-like ways and especially in their local contexts, groupings and actions. A Bostonian investment banker and an Amazonian hunter or an *erectus* sailor find their place and the role they occupy in society. These roles are not usually invented by the individual. They emerge or are blocked from emerging by a particular culture. One knows that there were no full-time *Homo erectus* musicians because there can be no such roles without an entire technology, social role and payment structure produced by society over time. And the structures and roles of the cultural-grammatical system into which we are born themselves emerge from the values and beliefs of a culture. In this sense, if we take culture as beliefs, knowledge and values, and society as roles and structural relationships between them, with members of society filling particular slots created by the culture, then at times it becomes easier to understand or at least visualise what people do as members of their culture.

Therefore, one may conceive of all the individuals of a society as 'fillers' for slots in a culture-grammar. One example is the college class-room. The fillers of the different classroom slots are easy – these are the students and the professor.

What kinds of roles and structures would an *erectus* society have had? Or what kind of roles and structures would another kind of primate society have? If we take an 'alpha-male' society of, say, gorillas, the typical social structure would be a silver-backed male (the alpha-male), pre-adult males and females, and females of mating age or beyond. In more complex gorilla societies there can be more than one silverback, but the typical arrangement is one silverback and many females and children. The male has a variety of duties, including decision-making for the group, resolution of conflicts, mating for the reproductive sur-vival of the group, deciding when the group should sleep and defending the group. *Erectus* societies would have had at least this level of organi-sation. In fact, as hunter-gatherers with *Homo* brains, they would have had a social structure simpler than, perhaps, but comparable to that of some modern hunter-gathers. Consider an Amazonian society such as the Pirahãs. That society will be manifested by its individuals and form larger subunits that will include families, men, children, adolescents, women and so forth. A different tribal society group might instead be broken down into more structured kinship hierarchies, including families, clans, lineages, or more professional specialisations.

To act together, a society must in some way share the intention that our individual actions produce a result of the group. Voting is arguably such an action. Participating in a classroom lecture is another. These are all actions in the grammar of culture in which each person occupies a role, alone or jointly. In the social organisation exemplified above, the students are the object, not the subject matter. We are describing their social roles in this moment in time relative to a particular teacher. Their roles may shift slightly with their next class. Certainly, students and teachers will change their roles at parties, at their homes and in their careers. Roles are like apparel, worn for specific situations.

When participants are from different cultures, as in the Treaty of Medicine Lodge example, they often assume that everyone shares a similar understanding of roles, structures and meanings of the joint act they are engaging in. But they rarely realise that each participant possesses a separate interpretation of their joint activity. In my view of the entire situation, this is what happened: the Comanches interpreted the promises made by the US government at the Medicine Lodge event as effective immediately and unconditionally. To them everyone speaking was a plenipotentiary representative of their people. The US negotiators, however, saw themselves as subordinates to Congress and perceived the Indians as a group that should accede to this greater authority. They understood the joint act of treaty-signing as entering into a conditional, time-delayed initial offer. (They also saw the Indians as inferior beings whose opinions and understanding mattered less.)

The societies of *Homo erectus* would have included criteria for membership in each community, the duties of each community member, relationships between members, such as children and adults, activity-planning and other needs.

Perceptions and the range of thought are shaped significantly by a cultural network. This turns out to mean in European societies that the dualism of Descartes and the mind-as-computer idea of Alan Turing represent the core of cognition. But this seems misguided.

Since the earliest days of artificial intelligence, eminent proponents of the idea that brains are computers have proposed, often quite emotionally, that of course machines can think. John McCarthy says the following: 'To ascribe certain *beliefs, knowledge, free will, intentions, consciousness, abilities,* or *wants* to a machine or computer program is legitimate when such an ascription expresses the same information about the machine that it expresses about a person.'[1]

But this kind of statement is built on a faulty understanding of beliefs as well as a faulty understanding of culture. And the personification of computers by attributing beliefs and so on to them one often hears suggested is too powerful. It could be extended in humorous, but no less valid, ways to circumstances no one would ascribe beliefs to. One could say that a thermostat believes it is too hot so it turns on the air-conditioning. Or that toes curl up because they believe it is warm. Or that plants turn towards the sun, because they believe they should. In fact, there are many cultures, the Pirahã's and Wari's, in which beliefs are regularly ascribed to animals, to clouds, to trees and so on as a convenient way of talking. But the tribes I have worked with don't usually mean these ascriptions literally.

Beliefs are states which occur when bodies (including brains) are directed towards something, from an idea to a plant. Beliefs are formed by the individual as she or he engages in language and culture.

As one contemplates *erectus* culture, values, beliefs and social roles, some subsidiary issues come to mind: the role and emergence of tools in any culture. How is one to characterise tools culturally, things that are used to aid individual cultural members in different tasks? Tools are dripping with cultural knowledge. One could even conceive of tools as congealed culture. Examples include physical tools, such as shovels, paintings, hats, pens, plates and food. But also non-physical tools are crucial. Perhaps humans' most important tool is language. In fact, culture itself is a tool.

The tool-like nature of language can be seen easily in its stories. Stories are used to exhort, to explain, to describe and so on, and each text is embedded in a context of dark matter. Stories, including books, are, of course, unlike physical tools in the sense that as linguistic devices they could in principle have revealed something about the dark matter from which they partially emerge, though generally very little is conveyed. And the reason for that is clear. People talk about what they assume their interlocutor does not know (but has the necessary background knowledge to understand). And tacit knowledge, or dark matter, which people are usually unaware of, is simply overlooked.

That language as a tool is also seen in the forms of stories. Consider a list of principles that anthropologist Marvin Harris provided to account for the Hindu rules governing defecation in Indian rural areas:

> A spot must be found not too far from the house.
> The spot must provide protection against being seen.
> It must offer an opportunity to see any one approaching.
> It should be near a source of water for washing.
> It should be upwind of unpleasant odours.
> It must not be in a field with growing crops.[2]

The first line uses the indefinite article 'a.' In the second line the definite article 'the' is used. From that point onward, 'spot' is pronominalised as 'it'. This is because of English conventions for keeping track of a topic through a discourse. The indefinite article indicates that the noun it modifies is new information. The definite shows that it is shared information. The pronoun reveals that it is topical. As the single word is referenced and re-referenced throughout the discourse its changing role and relationship to shared knowledge is marked with specific grammatical devices. This is shared but unspoken and largely ineffable knowledge to the non-specialist.

How does the understanding of culture promoted here compare to the wider understanding of culture in a society as a whole? It is common to hear about 'American culture', 'Western values', or even 'pan-human values' and so forth. According to the theory of dark matter and culture developed above these are perfectly sensible ideas, so long as we interpret them to mean 'overlapping values, rankings, roles and knowledge', rather than a complete homogeneity of (any notion of) culture throughout a given population. From laws to pronunciation, from architecture to music to sexual positions and body shape, the actions of individual humans as members of communities ('likers of Beethoven', 'eaters of haggis' and on and on) in conjunction with an individual's apperceptions and episodic memory – all are the products of overlapping dark matters.

Just so, values can produce in an individual or in a community a sense of *mission* – such as the Boers, the Zionists, the American frontiersmen and settlers who subscribed to manifest destiny, or the National Socialists who dreamed of a thousand-year Reich. This sense of mission and purpose is what many businesses are after today as the use of the term culture has been adopted by companies as 'what they are all about'. Did *erectus* communities have any sense of mission?

Although there are most certainly general principles of human behaviour and the formation of dark matter, the combination of individual apperceptions with exposure to mere subsets of larger value, knowledge and role networks means that no two people will be exactly alike in any way. And certainly no two cultures will be.

There are better examples of knowledge that is unspoken, though. Non-human animals present superior examples in some ways. These animals have beliefs, desires and emotions, learn complicated behaviours and ways of interacting with the world. Yet they lack language altogether and so, by definition, cannot talk about their knowledge. Almost all non-human animal knowledge is therefore dark matter. Most people wave their hands at these fascinating phenomena, sweeping them all under the label of 'instincts' rather than knowledge.

Dogs, humans and other animals go through an attachment period, are driven by emotions, learn tricks, learn to obey a range of commands, come to sense ownership/relationship/belonging to certain items in their environment and so on. My 140-pound Fila Brasileiro barks when even slight changes are added to the environment – a stack of books in a strange place, cushions from the sofa piled for cleaning, a new car in the driveway and so on. While my dog cannot 'tell' me about this in English, through her barking and body posture she communicates relatively well, though many of her actual feelings remain ineffable. Her dark matter in this sense has both 'communicable' (via actions and barking) and ineffable components, just as human dark matter does. *Erectus*, like *sapiens*, would have learned their languages through interaction with other members of their community, especially their mothers.

Still other cultural conventions include queuing. In an American store no matter how crowded, most people will, without being instructed, form a queue in front of the cash register. In some countries, without rigorous enforcement, such queuing will not occur – everyone will crowd around the cash register hoping to get served first. Queuing is thus a convention of some cultures but not others. And, as with all conventions, when we experience another culture, we will always be bothered by the absence of our culture's conventions. The reason is that conventions make life easier by requiring fewer decisions, by bringing a sense of the familiar to the foreign.

Societies depend on conventions to be able to function. It is likely that *erectus* communities began to develop conventions. Who speaks first when two people meet? How do children get food in the presence of adults? Who is the first one to depart the village on a new journey? Philosopher Ruth Millikan claims that conventions share a range of properties, such as being able to be reproduced, the need for a precedent before something can become a convention, usefulness in organising actions (such as forming lines at the box office instead of everyone crowding around at once).[3] She also notes that we can all violate conventions for different reasons and effects, just as Grice observed that we can flout conversational maxims. Millikan asserts that all people want, expect and seek conventions, such as using a bag to hold a seat in a waiting room.

The discussion of the importance of conventions and the individual to culture leads to an understanding of culture as the core of cognition. The argument is that without culture there can be no semantic understanding, no background, no tacit knowledge to undergird new thought.

Erectus societies had culture. From the very first, humans, with their larger brains and new experiences, built up values, knowledge and social roles that allowed them to wander the earth, sail the seas and build the first communities in the history of the earth. And from these cultures built more than 60,000 generations ago we emerged. Our debt to *Homo erectus* is inestimable. They were not cavemen. They were men, women and children, the first humans to speak and to live in culturally linked communities.

Conclusion

That is why it was called Babel – because there the Lord confused
the language of the whole world. From there the Lord scattered
them over the face of the whole earth.

Genesis 11:9

MORE THAN 60,000 GENERATIONS AGO, *Homo erectus* introduced
language into the world. Not merely another form of animal communication, language is an advanced form of cultural expression, based
on abilities unique to human cognition along with general principles
of structure for information transfer.

The core of language is the symbol, a combination of a culturally
agreed upon form with a culturally developed meaning. Human perceptual constraints and thinking limitations guide this process, but it
is largely the output of human societies, their values, their knowledge
and their social structures.

The symbol may have resulted from associating two objects by
mistake, such as a tree root confused with a serpent, or simply by regular
association of one thing in the world with another object or event, as
Pavlov's dog learned to associate food with the ringing of a bell. Once
this connection was made, humans began to use their symbols, each
one learning from the other. Since communication is an effort of the
entire being, gestures, intonation, the lungs, the mouth, the tongue, the
hands, body movements and even eyebrows were marshalled for use in
language, just as they are in much other animal communication. These
different components of our communicative effort in language would
have broken symbols down into smaller and smaller parts as they also
were used to build them into larger and larger units. Speech sounds,

words, sentences, grammatical affixes and tones all emerged from the initial invention of the symbol, with the invention being improved and spreading over time by total societal involvement, just as all other inventions are. Meaningless elements (sounds like 's', 'a' and 't') were combined to form meaningful items (such as the word 'sat') and duality of patterning emerged, itself leading next to three types of grammar. The first kind of grammar, G_1, is little more than symbols arranged in rows like beads on a string: 'Eat food. Man. Woman.' Or even, 'I see you. You see me?' The next language type, G_2, arranges symbols linearly (in a row), just like a G_1 grammar, and hierarchically – combining symbols inside of other symbols, just as many modern European languages do. The third type of grammar, G_3, does everything that the other types do, but with the added property of recursion, the ability to put one thing inside another thing of the same type without end. Language as a matryoshka doll. All three types of languages are still found in the world. All are fully functioning human languages appropriate for different cultural niches. *Homo erectus* communities spoke one or all of these types of grammars, in their far-flung outposts around the world.

Human languages change over time and cultures and speakers elaborate them in some places and simplify them in others. Contemporary languages are therefore different in their details from those of 2 million years ago. But the fact remains that 2 million years ago in Africa, a *Homo erectus* community began to share information among its members by means of language. They were the first to say, 'It's over there,' or, 'I am hungry.' Maybe the first to say, 'I love you.'

Erectus communities were unlike *sapiens* communities in many ways. But they were nevertheless societies of human beings discussing, deliberating, debating and denouncing, as they travelled the world and bequeathed to us their invention, language.

Each human alive enjoys their grammar and society because of the work, the discoveries and the intelligence of *Homo erectus*. Natural selection took those things that were most effective for human survival and improved the species until today humans live in the Age of Innovation, the Era of Culture, in the Kingdom of Speech.

Suggested Reading

Anderson, Michael L. *After Phrenology: Neural Reuse and the Interactive Brain*. Cambridge, MA: MIT Press, 2014.

Arbib, Michael A. 'From Monkey-Like Action Recognition to Human Language: An Evolutionary Framework for Neurolinguistics'. *Behavioral and Brain Sciences* 28(2), 2005: 105–124.

——. *How the Brain Got Language: The Mirror System Hypothesis*. Oxford University Press, 2012.

Barnard, Alan. *Genesis of Symbolic Thought*. Cambridge University Press, 2012.

Bednarik, R. G. 'Concept-Mediated Marking in the Lower Palaeolithic'. *Current Anthropology* 36, 1995: 605–634.

——. 'The "Australopithecine" Cobble from Makapansgat, South Africa'. *South African Archaeological Bulletin*, 53, 1998: 4–8.

——. 'Maritime Navigation in the Lower and Middle Palaeolithic'. Comptes Rendus de l'Académie des Sciences Paris, *Earth and Planetary Sciences*, 328, 1999: 559–563.

——. 'Seafaring in the Pleistocene'. *Cambridge Archaeological Journal*, 13(1), 2003: 41–66.

——. 'A Figurine from the African Acheulian'. *Current Anthropology*, 44(3), 2003: 405–413.

——. 'Middle Pleistocene Beads and Symbolism'. *Anthropos*, 100(2), 2005: 537–552.

——. 'Beads and the Origins of Symbolism'. *Time and Mind: The Journal of Archaeology, Consciousness and Culture* 1(3), 2008: 285–318.

——. 'On the Neuroscience of Rock Art Interpretation'. *Time and Mind: The Journal of Archaeology, Consciousness and Culture* 6(1), 2013: 37–40.

——. 'Exograms'. *Rock Art Research*, 31(1), 2014: 47–62.

——. 'Doing with Less: Hominin Brain Atrophy'. *HOMO – Journal of Comparative Human Biology*, 65, 2014: 433–449; doi: 10.1016/j.jchb.2014.06.001

——. 'Mind and Creativity of Hominins'. *SemiotiX: A Global Information Magazine*, February 2017.

Bedny, Marina, Hillary Richardson and Rebecca Saxe. '"Visual" Cortex Responds to Spoken Language in Blind Children'. *Journal of Neuroscience* 35(33), 2015: 11674 –11681.

Berent, Iris. *The Phonological Mind*. Cambridge University Press, 2013.

Berwick, Robert C. and Noam Chomsky. *Why Only Us?* Cambridge, MA: MIT Press, 2016.

Bolhuis, Johan J., Martin Everaert, Robert Berwick and Noam Chomsky. *Birdsong, Speech and Language: Exploring the Evolution of Mind and Brain*. Cambridge, MA: MIT Press, 2016.

Boyd, Robert and Peter J. Richerson. *Culture and the Evolutionary Process*. University of Chicago Press, 1988.

——. *The Origin and Evolution of Cultures*. Oxford University Press, 2005.

Brandom, Robert B. *Making it Explicit: Reasoning, Representing, and Discursive Commitment*. Cambridge, MA: Harvard University Press, 1998.

Bybee, Joan L. *Language, Usage and Cognition*. Cambridge University Press, 2010.

Cangelosi, Angelo. 'Evolution of Communication and Language Using Signals, Symbols and Words'. *IEEE Transactions on Evolutionary Computation* 5(2), 2001: 93–101.

Chomsky, Noam. 'Formal Properties of Grammars'. In R. Duncan Luce, Robert R. Bush and Eugene Galanter (eds), *Handbook of Mathematical Psychology*, vol. 2. New York: John Wiley, 1963, pp. 323–418.

——. *Language and Mind*, enlarged edn. New York: Harcourt Brace Jovanovich, 1972.

——. 'On Language and Culture'. In Wiktor Osiatyński (ed.), *Contrasts: Soviet and American Thinkers Discuss the Future*. New York: Macmillan, 1984, pp. 95–101.

——. *Knowledge of Language: Its Nature, Origin and Use*. New York: Praeger, 1986.

——. *The Minimalist Program*. Cambridge, MA: MIT Press, 1995.

——. 'Minimal Recursion: Exploring the Prospects'. In Tom Roeper and Margaret Speas (eds), *Recursion: Complexity in Cognition*. Cham: Springer International, 2014, pp. 1–15.

Corballis, Michael C. *From Hand to Mouth*. Princeton University Press, 2002.

——. 'Recursion, Language and Starlings'. *Cognitive Science* 31(4), 2007: 697–704.

De Ruiter, Jan P. and David Wilkins. 'The Synchronisation of Gesture and Speech in Dutch and Arrernte (an Australian Aboriginal Language)'. In S. Santi, I. Guaïtella, C. Cavé and G. Konopczynski (eds), *Oralité et Gestualité*. Paris: L'Hamattan, 1998, pp. 603–607.

Dediu, Dan and Steven C. Levinson. 'On the Antiquity of Language: The Reinterpretation of Neanderthal Linguistic Capacities and Its Consequences'. *Frontiers in Psychology*, 5 July 2013. doi:10.3389/fpsyg.2013.00397.

Diller, Karl C. and Rebecca L. Cann. 'Evidence Against a Genetic-Based Revolution in Language 50,000 Years Ago'. In R. Botha and C. Knight (eds), *The Cradle of Language*. New York: Oxford University Press, 2009, pp. 135–149.

Dunbar, Robin. *Grooming, Gossip and the Evolution of Language*. Cambridge, MA: Harvard University Press, 1998.

Efron, David. *Gesture and Environment*. New York: King's Crown Press, 1941.

Evans, Vyvyan. 'Beyond Words: How Language-Like is Emoji?' Oxford Dictionaries blog, 20 November 2015. http://blog.oxforddictionaries. com/2015/11/emoji-language/.

Everaert, Martin B., Marinus A. Huybregts, Noam Chomsky, Robert C. Berwick and Johan J. Bolhuis. 'Structures, Not Strings: Linguistics as Part of the Cognitive Sciences'. *Trends in Cognitive Sciences* 19(12), 2015: 729–743.

Everett, Caleb D. 'Evidence for Direct Geographic Influences on Linguistic Sounds: The Case of Ejectives'. *PLoS ONE* 8(6), 2014: e65275. doi:10.1371/journal.pone.0065275.

——. 'Climate, Vocal Folds and Tonal Languages' (with D. Blasi and S. Roberts). *Proceedings of the National Academy of Sciences of the United States of America* 112(5), 2015: 1322–1327.

Everett, Daniel L. 'Aspectos da Fonologia do Pirahã'. Master's thesis, Universidade Estadual de Campinas, 1979. http://ling.auf.net/lingbuzz/001715.

——. 'Phonetic Rarities in Pirahã'. *Journal of the International Phonetics Association* 12, 1982: 94–96.

——. 'A Lingua Pirahã e a Teoria da Sintaxe'. PhD dissertation, Universidade Estadual de Campinas, 1983. Published as *A Lingua Pirahã e a Teoria da Sintaxe*, Campinas: Editora da UNICAMP, 1992.

——. 'Syllable Weight, Sloppy Phonemes and Channels in Pirahã Discourse'. In Mary Niepokuj et al. (eds.), *Proceedings of the Eleventh Annual Meeting Berkeley Linguistics Society*. Berkeley Linguistics Society, 1985, pp. 408–416.

——. 'Pirahã'. In Desmond Derbyshire and Geoffrey Pullum (eds), *Handbook of Amazonian Languages I*. Berlin: de Gruyter, 1986, pp. 200–326.

——. 'On Metrical Constituent Structure in Pirahã Phonology'. *Natural Language and Linguistic Theory* 6, 1988: 207–246.

——. 'The Sentential Divide in Language and Cognition: Pragmatics of Word Order Flexibility and Related Issues'. *Journal of Pragmatics and Cognition* 2(1), 1994: 131–166.

——. 'Monolingual Field Research'. In Paul Newman and Martha Ratliff (eds), *Fieldlinguistics*. Cambridge University Press, 2001, pp. 166–188.

——. 'Coherent Fieldwork'. In P. van Sterkenburg (ed.), *Linguistics Today*, Amsterdam: John Benjamins, 2004, pp. 141–162.

——. 'Periphrastic Pronouns in Wari'. *International Journal of American Linguistics* 71(3), 2005: 303–326.

——. 'Cultural Constraints on Grammar and Cognition in Pirahã: Another Look at the Design Features of Human Language'. *Current Anthropology* 76, 2005: 621–646.

——. *Don't Sleep, There Are Snakes: Life and Language in the Amazonian Jungle.* New York: Pantheon, 2008.

——. 'Wari' Intentional State Construction Predicates'. In Robert Van Valin (ed.), *Investigations of the Syntax-Semantics-Pragmatics Interface.* Amsterdam: John Benjamins, 2009, pp. 381–409.

——. 'Pirahã Culture and Grammar: A Response to Some Criticisms'. *Language* 85(2), 2009: 405–442.

——. 'You Drink. You Drive. You Go to Jail. Where's Recursion?' Paper presented at the 2009 University of Massachusetts Conference on Recursion. http://ling. auf.net/lingbuzz/001141.

——. 'The Shrinking Chomskyan Corner in Linguistics'. Response to the criticisms Nevins, Pesetsky and Rodrigues raise against various papers of Everett on Pirahã's unusual features, published in *Language* 85, http://ling.auf. net/lingbuzz/000994/current.pdf, 2010.

——. *Language: The Cultural Tool.* New York: Pantheon Books, 2012.

——. 'What Does Pirahã Have to Teach about Human Language and the Mind?' *WIREs Cognitive Science.* doi:10.1002/wcs.1195, 2012.

——. 'A Reconsideration of the Reification of Linguistics'. Paper presented at The Cognitive Revolution, 60 Years at the British Academy, London, 2013.

——. 'The State of Whose Art?' Reply to Nick Enfield's review of *Language: The Cultural Tool* in *Journal of the Royal Anthropological Institute* 19(1), 2013.

——. 'Concentric Circles of Attachment in Pirahã: A Brief Survey'. In Heidi Keller and Hiltrud Otto (eds), *Different Faces of Attachment: Cultural Variations of a Universal Human Need.* Cambridge University Press, 2014, pp. 169–186.

——. 'The Role of Culture in the Emergence of Language'. In Brian MacWhinney and William O'Grady (eds), *The Handbook of Language Emergence.* Hoboken, NJ: Wiley-Blackwell, 2014, pp. 354–376.

——. *Dark Matter of the Mind: The Culturally Articulated Unconscious.* University of Chicago Press, 2016.

Everett, Daniel L. and Keren Everett. 'On the Relevance of Syllable Onsets to Stress Placement'. *Linguistic Inquiry* 15, 1984: 705–711.

Fitch, W. Tecumseh. *The Evolution of Language.* Cambridge University Press, 2010.

Floyd, Simeon. 'Modally Hybrid Grammar? Celestial Pointing for Time-of-Day Reference in Nheengatú'. *Language* 92(1), 2016: 31–64. doi:10.1353/lan.2016.001.

Freyd, Jennifer. 'Shareability: The Social Psychology of Epistemology'. *Cognitive Science* 7, 1983: 191–210.

Fuentes, Augustin. 'The Extended Evolutionary Synthesis, Ethnography and the Human Niche: Toward an Integrated Anthropology'. *Current Anthropology* 57, supp. 13, June 2016.

Futrell, Richard, Laura Stearns, Steven T. Piantadosi, Daniel L. Everett and Edward Gibson. 'A Corpus Investigation of Syntactic Embedding in Pirahã'. *PLoS ONE*, 11(3), 2016: e0145289. doi:10.1371/journal.pone.0145289.

Gil, David. 'The Structure of Riau Indonesian'. *Nordic Journal of Linguistics* 17, 1994: 179–200.

Goldberg, Adele. *Constructions: A Construction Approach to Argument Structure.* University of Chicago Press, 1995.

——. *Constructions at Work: The Nature of Generalisation in Language.* Oxford University Press, 2006.

Grice, Paul. *Studies in the Way of Words.* Cambridge, MA: Harvard University Press, 1991.

Harris, Marvin. *Cultural Anthropology.* Boston: Allyn & Bacon, 1999.

——. *Cultural Materialism: The Struggle for a Science of Culture.* Walnut Creek, CA: Altamira, 2001.

Harris, Zellig. *Methods in Structural Linguistics.* University of Chicago Press, 1951.

Hauser, Marc, Noam Chomsky and Tecumseh Fitch. 'The Faculty of Language: What Is It, Who Has It, How Did It Evolve?' *Science* 298, 2002: 1569–1579.

Heckenberger, Michael J., J. Christian Russell, Carlos Fausto, Joshua R. Toney, Morgan J. Schmidt, Edithe Pereira, Bruna Franchetto, Afukaka Kuikuro. 'Pre-Columbian Urbanism, Anthropogenic Landscapes and the Future of the Amazon'. *Science* 321, 2008: 1214–1217.

Hickok, Gregory. *The Myth of Mirror Neurons: The Real Neuroscience of Communication and Cognition*, New York: W. W. Norton, 2014.

Hobbs, Jerry R. 'Deep Lexical Semantics'. In *Proceedings of the Ninth International Conference on Intelligent Text Processing and Computational Linguistics* (CICLing-2008), Haifa, Israel, February 2008.

Hockett, Charles. 'The Origin of Language'. *Scientific American* 203, 1960: 89–97.

Hopper, Paul. 'Emergent Grammar and the A Priori Grammar Postulate'. In Deborah Tannen (ed.), *Linguistics in Context: Connecting Observation and Understanding.* New York: Ablex, 1988.

Hurford, James R. *The Origins of Meaning: Language in the Light of Evolution.* Oxford University Press, 2011.

Jackendoff, Ray. *Foundations of Language: Brain, Meaning, Grammar, Evolution.* Oxford University Press, 2003.

Jackendoff, Ray and Eva Wittenberg after 'What You Can Say Without Syntax: A Hierarchy of Grammatical Complexity'. In Frederick J. Newmeyer and Laurel B. Preston (eds), *Measuring Grammatical Complexity*, Oxford University Press, 2014, ch 4; doi:10.1093/acprof:oso/9780199685301.003.0004.

Karlsson, Fred. 'Origin and Maintenance of Clausal Embedding Complexity'. In
    Geoffrey Sampson, David Gil and Peter Trudgill (eds), *Language Complexity as
    an Evolving Variable*. Oxford University Press, 2009, pp. 192–202.
Keller, Timothy A. and Marcel Adam Just. 'Altering Cortical Connectivity:
    Remediation-Induced Changes in the White Matter of Poor Readers'. *Neuron*
    64(5), 2009: 624–631.
Kendon, Adam. *Gesture: Visible Action as Utterance*. Cambridge University Press,
    2004.
Kinsella, Anna R. *Language Evolution and Syntactic Theory*. Cambridge
    University Press, 2009.
Kirby, Simon, Hannah Cornish and Kenny Smith. 'Cumulative Cultural Evolution
    in the Laboratory: An Experimental Approach to the Origins of Structure in
    Human Language'. *Proceedings of the National Academy of Sciences of the United
    States of America* 105(31), 2008: 10681–10686. doi:10.1073/pnas.0707835105.
Kirby, Simon, Mike Dowman and Thomas L. Griffiths. 'Innateness and Culture in
    the Evolution of Language'. *Proceedings of the National Academy of Sciences of
    the United States of America* 104(12), 2007. doi:10.1073/pnas.0608222104.
Labov, William. *Principles of Linguistic Change*, vol. 3: *Cognitive and Cultural
    Factors*. Oxford: Wiley-Blackwell, 2010.
LeDoux, Joseph. *Anxious: Using the Brain to Understand and Treat Fear and
    Anxiety*. New York: Viking, 2015.
Levinson, Stephen C. 'On the Human "Interaction Engine"'. In Nick J. Enfield and
    Stephen C. Levinson (eds), *Roots of Human Sociality: Culture, Cognition and
    Interaction*. New York: Berg, 2006, pp. 399–460.
——. 'Recursion in Pragmatics'. *Language*, 89(1), 2013: 149–162.
Levinson, Stephen C. and Pierre Jaisson. *Evolution and Culture: A Fyssen
    Foundation Symposium*. Cambridge, MA: MIT Press, 2005.
Levinson, Stephen C. and Asifa Majid. 'Differential Ineffability and the Senses'.
    Mind and Language 29(4), 2014: 407–427.
Lieberman, Philip. 'The Evolution of Human Speech: Its Anatomical and Neural
    Bases'. *Current Anthropology* 48(1), 2007: 39–66.
——. *The Unpredictable Species: What Makes Humans Unique*. Princeton
    University Press, 2013.
Longacre, Robert. *Grammar Discovery Procedures: A Field Manual*. The Hague:
    Mouton & Co., 1964.
Luuk, Erkki. 'The Structure and Evolution of Symbols'. *New Ideas in Psychology*
    31(2), 2013: 87–97.
Luuk, Erkki and Hendrik Luuk. 'The Evolution of Syntax: Signs, Concatenation
    and Embedding'. *Cognitive Systems Research* 27, 2014: 1–10. doi:10.1016/j.
    cogsys.2013.01.00.
Lyell, Charles. *Principles of Geology*. London: John Murray, 1833.

MacWhinney, Brian. 'A Unified Model of Language Acquisition'. In J. Kroll and A. de Groot (eds), *Handbook of Bilingualism: Psycholinguistic Approaches*. Oxford University Press, 2004, pp. 49–67.

——. 'Emergentism – Use Often and With Care'. *Applied Linguistics* 27(4), 2006: 729–740. doi:10.1093/applin/aml035.

MacWhinney, Brian and William O'Grady (eds). *The Handbook of Language Emergence*. Hoboken, NJ: Wiley-Blackwell, 2016.

McNeill, David. *Hand and Mind: What Gestures Reveal About Thought*. University of Chicago Press, 1992.

——. *Gesture and Thought*. University of Chicago Press, 2005.

——. *How Language Began: Gesture and Speech in Human Evolution*. Cambridge University Press, 2012.

McNeill, David (ed.). *Language and Gesture*. Cambridge University Press, 2000.

Mead, George Herbert. *Mind, Self and Society*. University of Chicago Press, 2015.

Morgan. T. J. H., N. T. Uomini, L. E. Rendell, L. Chouinard-Thuly, S. E. Street, H. M. Lewis, C. P. Cross, C. Evans, R. Kearney, I. de la Torre, A. Whiten and K. N. Laland. 'Experimental Evidence for the Co-Evolution of Hominin Tool-Making Teaching and Language'. *Nature Communications*, 6, 2015: 6029. doi: 10.1038/ncomms7029.

Müller, R. A. and S. Basho. 'Are Nonlinguistic Functions in "Broca's Area" Prerequisites for Language Acquisition? FMRI Findings from an Ontogenetic Viewpoint'. *Brain and Language* 89(2), 2004: 329–336.

Panksepp, Jaak and Lucy Biven. *The Archaeology of Mind: Neuroevolutionary Origins of Human Emotions*. New York: W. W. Norton, 2012.

Peirce, C. S. *Semiotics and Significs*, ed. Charles Hardwick. Bloomington: Indiana University Press, 1977.

——. *The Essential Peirce*, vol. 1: *Selected Philosophical Writings (1867–1893)*. Bloomington: Indiana University Press, 1992.

——. *The Essential Peirce*, vol. 2: *Selected Philosophical Writings, 1893–1913*. Bloomington: Indiana University Press, 1998.

Pepperberg, Irene M. 'Evolution of Communication and Language: Insights from Parrots and Songbirds'. In Maggie M. Tallerman and Katherine R. Gibson (eds), *The Oxford Handbook of Language Evolution*. Oxford University Press, 2012, pp. 109–119.

Piantadosi, Steven T., Harry Tily and Edward Gibson. 'The Communicative Function of Ambiguity in Language'. *Cognition* 122(3), 2012: 280–291.

Pierrehumbert, Janet and Julia Hirschberg. 'The Meaning of Intonational Contours in the Interpretation of Discourse'. In P. R. Cohen, J. Morgan and M. E. Pollock (eds), *Intentions in Communication*. Cambridge, MA: MIT Press, 1990, pp. 271–311.

Pike, Kenneth L. *Language in Relation to a Unified Theory of the Structure of Human Behavior*, 2nd rev. edn. The Hague: Mouton & Co., 1967.

Rizzolatti, Giacomo and Michael A. Arbib. 'Language Within Our Grasp'. *Trends Neuroscience* 21, 1998: 188–194.

Rosenbaum, David A. *It's a Jungle in There: How Competition and Cooperation in the Brain Shape the Mind*. Oxford University Press, 2014.

Safina, Carl. *Beyond Words: What Animals Think and Feel*. New York: Henry Holt, 2015.

Saussure, Ferdinand. *A Course in General Linguistics*. Chicago, IL: Open Court Publishing, 1983.

Searle, John. 'Chomsky's Revolution in Linguistics'. *New York Review of Books*, 29 June 1972. [Reprinted in Gilbert Harman (ed.), *On Noam Chomsky: Critical Essays*. Garden City, NY: Anchor Books, 1974.]

Selkirk, E. 'On the Major Class Features and Syllable Theory'. In M. Aronoff and R. T. Oehrle (eds), *Language Sound Structure: Studies in Phonology*. Cambridge, MA: MIT Press, 1984, pp. 107–136.

Sereno, Martin I. 'Origin of Symbol-Using Systems: Speech, But Not Sign, Without the Semantic Urge'. *Philosophical Transactions of the Royal Society B* 369(1651), 2013: 20130303. doi:10.1098/rstb.2013.0303.

Shannon, Claude E. 'A Mathematical Theory of Communication'. *Bell System Technical Journal* 27, 1948: 379–423, 623–656.

Silverstein, Michael. 'Indexical Order and the Dialectics of Sociolinguistic Life'. *Language and Communication* 23, 2003: 193–229.

——. 'Cultural Concepts and the Language-Culture Nexus'. *Current Anthropology* 45(5), 2004: 621–652.

Silverstein, Michael and Greg Urban, eds. *Natural Histories of Discourse*. University of Chicago Press, 1996.

Simon, Herbert A. 'The Architecture of Complexity'. *Proceedings of the American Philosophical Society* 106(6), 1962: 467–482.

Slater, Peter. 'Bird Song and Language'. In Maggie M. Tallerman and Katherine R. Gibson (eds), *The Oxford Handbook of Language Evolution*. Oxford University Press, 2012, pp. 96–101.

Sperber, Dan and Deirdre Wilson. *Relevance: Communication and Cognition*. Hoboken, NJ: Wiley-Blackwell, 1996.

Steedman, Mark. *The Syntactic Process*. Cambridge, MA: Bradford Books/MIT Press, 2001.

Steels, L. 'The Emergence and Evolution of Linguistic Structure: From Lexical to Grammatical Communication Systems'. *Connection Science* 17, 2005: 213–230.

Sterelny, Kim. *Thought in a Hostile World: The Evolution of Human Cognition*. Hoboken, NJ: Wiley-Blackwell, 2008.

——. *The Evolved Apprentice: How Evolution Made Humans Unique*. Cambridge, MA: Bradford Books/MIT Press, 2014.

Tallerman, Maggie and Kathleen R. Gibson (eds). *The Oxford Handbook of Language Evolution*. Oxford University Press, 2012.

Tattersall, Ian. *Masters of the Planet: The Search for Our Human Origins*. New York: St. Martin's Press, 2012.

Thomason, Sarah Grey and Terrence Kaufman. *Language Contact, Creolization and Genetic Linguistics*. Berkeley: University of California Press, 1992.

Thompson, B., Kirby, S. and Smith, K. 'Culture Shapes the Evolution of Cognition'. *Proceedings of the National Academy of Sciences of the United States of America* 113(16), 2016: 4530–4535. doi:10.1073/pnas.1523631113.

Tomasello, Michael. *The Cultural Origins of Human Cognition*. Cambridge, MA: Harvard University Press, 2001.

——. *Constructing a Language: A Usage-Based Theory of Language Acquisition*. Cambridge, MA: Harvard University Press, 2005.

——. *Origins of Human Communication*. Cambridge, MA: Bradford Books/MIT Press, 2010.

——. *A Natural History of Human Thinking*. Cambridge, MA: Harvard University Press, 2014.

Urban, Greg. 'Metasignaling and Language Origins'. *American Anthropologist*, n.s. 104(1), 2002: 233–246.

Van Valin, Robert D. and Randy LaPolla. *Syntax: Structure, Meaning and Function*. Cambridge University Press, 1997.

Vygotsky, Lev S. *Mind in Society: The Development of Higher Psychological Processes*, ed. Michael Cole. Cambridge, MA: Harvard University Press, 1978.

Weinreich, Uriel, William Labov and Marvin I. Herzog. 'Empirical Foundations for a Theory of Language Change'. In W. Lehmann and Y. Malkiel (eds), *Directions for Historical Linguistics*. Austin: University of Texas Press, 1968, pp. 95–189.

Wilkins, David P. 'Spatial Deixis in Arrernte Speech and Gesture: On the Analysis of a Species of Composite Signal as Used by a Central Australian Aboriginal Group'. Paper 6 in Elisabeth André, Massimo Poesio and Hannes Rieser (eds), *Proceedings of the Workshop on Deixis, Demonstration and Deictic Belief in Multimedia Contexts, held on occasion of ESSLLI XI*, pp. 31–45. Workshop held in the section 'Language and Computation' as part of the Eleventh European Summer School in Logic, Language and Information, 9–20 August 1999, Utrecht, The Netherlands.

Notes

Preface

1. The unique neuronal density of the human brain is explained clearly by Suzana Herculano-Houzel, in her *The Human Advantage: A New Understanding of How Our Brain Became Remarkable* (Cambridge, MA: MIT Press, 2016).

Chapter 1: Rise of the Hominins

1. This is all discussed engagingly in Siddhartha Mukherjee's book, *The Gene: An Intimate History* (New York: Scribner, 2016).

Chapter 2: The Fossil Hunters

1. Daniel Lieberman's *The Evolution of the Human Head* (Cambridge, MA: The Belknap Press of Harvard University Press, 2011) is an excellent discussion of the evolution of the human head and its implications for human cognition. A good deal of the discussion of this chapter comes directly from my own *Language: The Cultural Tool* (London/New York: Profile/Vintage, 2012).
2. This is discussed in two relatively recent books, *Fire: The Spark that Ignited Human Evolution*, by Frances D. Burton (Albuquerque: University of New Mexico Press) and Richard Wrangham's *Catching Fire: How Cooking Made Us Human* (New York: Basic Books, 2009).
3. Platforms are discussed at length in my *Language: The Cultural Tool.*
4. Robert W. Lurz, *Mindreading Animals: The Debate Over What Animals Know About Other Minds* (Cambridge, MA: MIT Press, 2011).
5. To cite just a few, Robert Lurz's *Mindreading Animals*; Sue Taylor Parker, Robert W. Mitchell and H. Lyn Miles (eds), *The Mentalities of Gorillas and Orangutans: Comparative Perspectives* (Cambridge University Press, 2006); Daria Maestripieri's edited volume *Primate Psychology* (Cambridge, MA: Harvard University Press, 2005); *Beyond Words: What Animals Think and Feel* by Carl Safina (New York: Holt, 2015; this book is designed for the general, non-specialist reader); and *The Cultural Lives of Whales and Dolphins* by Hal Whitehead and Luke Rendell (University of Chicago Press, 2014).

6. Paul M. Churchland, *Plato's Camera: How the Physical Brain Captures a Landscape of Abstract Universals* (Cambridge, MA: MIT Press, 2013), p. 22.

Chapter 4: Everyone Speaks Languages of Signs

1. Taken from my *Dark Matter of the Mind: The Culturally Articulated Unconscious* (University of Chicago Press, 2016).

2. Searle's 1972 review in the *New York Review of Books* of Chomsky's revolution: www.nybooks.com/articles/1972/06/29/a-special-supplement-chomskys-revolution-in-lingui/.

3. In *Stone Tools in Human Evolution: Behavioral Differences Among Technological Primates* (Cambridge University Press, 2016), for example, palaeoanthropologist John Shea discusses the links between tools and language.

4. This is taken from Johan J. Bolhuis and Martin Everaert (eds), *Birdsong, Speech, and Language: Exploring the Evolution of Mind and Brain* (Cambridge, MA: MIT Press, 2015), p. 729.

5. S. T. Piantadosi, H. Tily and E. Gibson, 'The Communicative Function of Ambiguity in Language', *Cognition* 122(3), 2012: 280–291; doi: 10.1016/j.cognition.2011.10.004.

6. Such as Michael Anderson in his 2014 book, *After Phrenology: Neural Reuse and the Interactive Brain* (Cambridge, MA: MIT Press), and Stanislas Dehaene in *Reading in the Brain* (New York: Viking, 2009).

7. See, for example, Robert C. Berwick, and Noam Chomsky, *Why Only Us? Language and Evolution* (Cambridge, MA: MIT Press, 2016); Martin B. Everaert et al., 'Structures, Not Strings: Linguistics as Part of the Cognitive Sciences', *Trends in Cognitive Sciences* 19(12), 2015: 729–743, a prolegomenon which, I hope, complements other empirical work: see Maggie Tallerman and Kathleen R. Gibson (eds.), *The Oxford Handbook of Language Evolution* (Oxford University Press, 2012); B. Thompson, S. Kirby and K. Smith, 'Culture Shapes the Evolution of Cognition', *Proceedings of the National Academy of Sciences of the United States of America* 113(16), 2016: 4530–4535; James R. Hurford, *The Origins of Meaning: Language in the Light of Evolution* (Oxford University Press, 2011).

8. These terms come from the work of American anthropological linguist, Kenneth L. Pike, *Language in Relation to a Unified Theory of the Structure of Human Behavior*, 2nd rev. edn (The Hague/Paris: Mouton & Co., 1967).

9. www.zmescience.com/science/archaeology/homo-erectus-shell-04122014/.

10. T. J. H. Morgan et al., 'Experimental Evidence for the Co-Evolution of Hominin Tool-Making, Teaching and Language', *Nature Communications* 6, 2015: 6029; doi: 10.1038/ncomms7029.

11. Robert Boyd and Peter Richerson are the leaders in the discussion of the roles of imitation vs innovation in cultural evolution and are the authors of many books about this tension. For example, see their *The Origin and Evolution of Cultures* (Oxford University Press, 2005) or *Culture and the Evolutionary Process* (University of Chicago Press, 1988).

12. The term 'satisficing' comes from the work of Nobel-prize-winning economist Herbert Simon in 1962. See, for example, 'The Architecture of Complexity', *Proceedings of the American Philosophical Society* 106(6): 467–482, and his 1947 book *Administrative Behavior: A Study of Decision-Making Processes in Administrative Organization* (New York: Macmillan).

13. Greg Urban, 'Metasignaling and Language Origins', *American Anthropologist*, New Series, 104(1), 2002: 233–246.

14. 'It is not likely that there was any single mutation causing the origin of language, or even speech, as seen by the complex relationship between FOXP2 and CNTNAP2 and by the fact that FOXP2 regulates several hundred genes, including many that have non-language related functions ...' Karl C. Diller and Rebecca L. Cann, 'The Innateness of Language: A View from Genetics', in Andrew D. M. Smith, Marieke Schouwstra, Bart de Boer and Kenny Smith (eds), *Proceedings of the 8th International Conference on the Evolution of Language* (Singapore: World Scientific, 2010), pp. 107–115.

15. Those interested in an analysis of Pirahã lacking grammar should consult Richard Futrell et al., 'A Corpus Investigation of Syntactic Embedding in Pirahã', at http://journals.plos.org/plosone/article?id=10.1371/journal.pone.0145289.

16. Caleb Everett has established this in extensive research on the interaction of climate, altitude and humidity on human sound systems, phonologies. (See Suggested Reading section.)

Chapter 5: Humans Get a Better Brain

1. Ralph L. Holloway, D. Broadfield and M. Yuan, *The Human Fossil Record*, vol. 3: *Brain Endocasts: The Paleoneurological Evidence* (Hoboken, NJ: John Wiley & Sons, 2004).

2. William R. Leonard, J. Josh Snodgrass and Marcia L. Robertson, 'Evolutionary Perspectives on Fat Ingestion and Metabolism in Humans', in J. P. Montmayeur and J. le Coutre (eds), *Fat Detection: Taste, Texture, and Post Ingestive Effects* (Boca Raton, FL: CRC Press/Taylor & Francis, 2010), chapter 1; www.ncbi.nlm.nih.gov/books/NBK53561/.

3. According to Indiana University palaeoneurologist Thomas Schoenemann in 'Evolution of the Size and Functional Areas of the Human Brain', *Annual*

Review of Anthropology 35, 2006: 379–406; www.indiana.edu/~brainevo/publications/annurev.anthro.35.pdf.

4. P. Tom Schoenemann, 'The Meaning of Brain Size: The Evolution of Conceptual Complexity', in Kathy Schick, Douglas Broadfield, Nicholas Toth and Michael Yuan (eds), *The Human Brain Evolving: Paleoneurological Studies in Honor of Ralph L. Holloway* (Gosport, IN: Stone Age Institute Press, 2010), pp. 37–50.

5. Mark Grabowski, 'Bigger Brains Led to Bigger Bodies?: The Correlated Evolution of Human Brain and Body Size', *Current Anthropology* 57(2), 2016: 174; doi: 10.1086/685655.

6. In an entertaining book, Falk provides the best popular account of palaeoneurology. *The Fossil Chronicles: How Two Controversial Discoveries Changed Our View of Human Evolution* (Berkeley, University of California Press, 2012).

Chapter 6: How the Brain Makes Language Possible

1. Much of the material at the beginning of this chapter is taken from my *Language: The Cultural Tool.*

2. Philip Lieberman, *Human Language and our Reptilian Brain: The Subcortical Bases of Speech, Syntax, and Thought* (Cambridge, MA: Harvard University Press, 2000).

3. D. M. Tucker, G. A. Frishkoff and P. Luu, 'Microgenesis of Language', in Brigette Stemmer and Harry A. Whitaker, eds), *Handbook of the Neuroscience of Language* (London: Elsevier, 2008), pp. 45–56.

4. Jeffrey Elman et al., *Rethinking Innateness: A Connectionist Perspective on Development* (Cambridge, MA: MIT Press, 1996), p. 241.

5. '… the observation that the term "Broca's region" (and that of "Wernicke's region") is not consistently used in the literature should come as no surprise. This inconsistency is not just a problem of nomenclature; rather, it is a conceptual one.' Katrin Amunts, 'Architectonic Language Research', in Brigitte Stemmer and Harry A. Whitaker (eds), *Handbook of the Neuroscience of Language* (London: Elsevier, 2008), pp. 33–44.

6. Ibid.

7. Ned T. Sahin, Steven Pinker, Sydney S. Cash, Donald Schomer and Eric Halgren. 'Sequential Processing of Lexical, Grammatical, and Phonological Information Within Broca's Area', *Science* 326(5951), 2009: 445–449; doi: 10.1126/science.1174481.

8. Miguel Nicolelis and Ronald Cicurel, *The Relativistic Brain: How it Works and Why it Cannot be Simulated by a Turing Machine* (Durham, NC: Kios Press, 2015).

9. Marina Bedny, Hilary Richardson and Rebecca Saxe, '"Visual" Cortex Responds to Spoken Language in Blind Children', *Journal of Neuroscience* 35(33), 2015: 11674–11681; doi: 10.1523/JNEUROSCI.0634-15.2015.

10. Evelina Fedorenko, 'The Role of Domain-General Cognitive Control in Language Comprehension', *Frontiers in Psychology* 5, 2014: 335.

11. Larry Swanson, *Brain Architecture: Understanding the Basic Plan* (Oxford University Press, 2011), p. 11.

12. See Lieberman, *The Evolution of the Human Head.*

13. Berwick and Chomsky, *Why Only Us?*

14. Pierre Perruchet and Arnaud Rey, 'Does the Mastery of Center-Embedded Linguistic Structures Distinguish Humans from Nonhuman Primates?' *Psychonomic Bulletin & Review* 12(2), 2005: 307–313.

15. http://itre.cis.upenn.edu/~myl/languagelog/archives/000434.html.

16. Noam Chomsky, *Cartesian Linguistics: A Chapter in the History of Rationalist Thought* (New York: Harper & Row, 1966).

17. The best history of many of these theories of the human mind in my opinion is the magisterial, two-volume work of Margaret A. Boden, *Mind as Machine: A History of Cognitive Science* (Oxford University Press, 2006). Another important history of studies of the mind is Willem J. M. Levelt's *A History of Psycholinguistics: The Pre-Chomskyan Era* (Oxford University Press, 2014).

18. Ralph Holloway, 'Brain Fossils: Endocasts', in L. R. Squire (ed.), *Encyclopedia of Neuroscience* (London: Academic Press, 2009), vol. 2, pp. 353–361.

Chapter 7: When the Brain Goes Wrong

1. Yves Turgeon and Joël Macoir, 'Classical and Contemporary Assessment of Aphasia and Acquired Disorders of Language', in Brigette Stemmer and Harry A. Whitaker (eds), *Handbook of the Neuroscience of Language* (London: Elsevier, 2008), pp. 3–11.

2. Michael Ullman and Elizabeth Pierpont, 'Specific Language Impairment is Not Specific to Language: The Procedural Deficit Hypothesis', *Cortex* 41(3), 2005: 399–433.

3. Ibid.

4. D. V. M. Bishop and M. E. Hayiou-Thomas, 'Heritability of Specific Language Impairment Depends on Diagnostic Criteria', *Genes, Brains, and Behavior* 7(3), 2008: 365–372; doi: 10.1111/j.1601-183X.2007.00360.xPMCID: PMC2324210.

5. Turgeon and Macoir, 'Classical and Contemporary Assessment', p. 5.

6. Edward Gibson, Chaleece Sandberg, Evelina Fedorenko, Leon Bergen and Swathi Kiran, 'A Rational Inference Approach to Aphasic Language

Comprehension', *Aphasiology* 30(11), 2015: 1341–1360; doi:10.1080/02687038. 2015.1111994.

7. Richard Griffin and Daniel Dennett, 'What Does the Study of Autism Tell Us About the Craft of Folk Psychology?', in T. Striano and V. Reid (eds), *Social Cognition: Development, Neuroscience, and Autism* (Hoboken, NJ: Wiley-Blackwell, 2008), pp. 254–280.

8. In Jacob A. Burack and Tony Charman, *The Development of Autism: Perspectives From Theory and Research* (New York and London: Routledge, 2015).

Chapter 8: Talking with Tongues

1. Philip Lieberman, 'Old-Time Linguistic Theories', *Cortex* 44, 2008: 218–226.

2. W. Tecumseh Fitch, Bart de Boer, Neil Mathur and Asif A. Ghazanfar, 'Monkey Vocal Tracts Are Speech-Ready', *Science Advances* 2(12), 2016; http://advances.sciencemag.org/content/2/12/e1600723; doi: 10.1126/sciadv.1600723.

3. These criticisms are not original with me. I have taken them almost exactly from an email by phonetician Caleb D. Everett of the University of Miami (the last name is not a coincidence).

4. Luigi Capasso, Elisabetta Michetti and Ruggero D'Anastasio, 'A Homo Erectus Hyoid Bone: Possible Implications for the Origin of the Human Capability for Speech', *Collegium antropologicum* 32(4), 2008: 1007–1011.

5. For a fuller account of the evolution and essential properties of hominid speech, I refer the reader to Philip Lieberman's *Toward an Evolutionary Biology of Language* (Cambridge, MA: The Belknap Press of Harvard University Press, 2006), from which I take much of the following material.

6. Ibid.

7. The following paragraphs borrow considerably from my *Language: The Cultural Tool*.

8. Paraphrased from Lieberman's *Toward an Evolutionary Biology of Language*.

Chapter 9: Where Grammar Came From

1. In his famous article, 'The Origin of Speech', *Scientific American* 203, 1960: 88–111.

2. Richard Futrell, et. al., 'A Corpus Investigation of Syntactic Embedding in Pirahã', *PLoS ONE* 11(3), 2016: e0145289; doi:10.1371/journal.pone.0145289, argue that there exist modern human languages that come in lower in the Chomsky hierarchy than Chomsky would have predicted.

3. Fred Karlsson, 'Origin and Maintenance of Clausal Embedding Complexity', in Geoffrey Sampson, David Gil and Peter Trudgill (eds), *Language*

Complexity as an Evolving Variable (Oxford University Press, 2009), pp. 192–202, explains his notation thus:

> 'I' stands for initial clausal embedding, 'C' for clausal centre-embedding, 'F' for final clausal embedding, and the raised exponent expresses the maximal degree of embedding of a sentence, e. g. I-2 is double initial embedding as in sentence (6). Expressions like C-2 indicate type and embedding depth of individual clauses; e. g. C-2 is a centre-embedded clause at depth 2.

4. Donald Davidson, 'On Saying That', *Synthese* 19, 1968: 130–146.
5. Searle's 1972 review in the *New York Review of Books* of Chomsky's revolution: www.nybooks.com/articles/1972/06/29/a-special-supplement-chomskys-revolution-in-lingui/.

Chapter 10: Talking with the Hands

1. I attempt to develop just such a theory in *Dark Matter of the Mind: The Culturally Articulated Unconscious*, from which a good deal of the material in this chapter is taken.
2. Pike, *Language in Relation to a Unified Theory of the Structure of Human Behavior*.
3. www.nytimes.com/2013/07/01/world/europe/when-italians-chat-hands-and-fingers-do-the-talking.html.
4. David McNeill, *Gesture and Thought* (University of Chicago Press), 2005, p. 117.

Chapter 12: Communities and Communication

1. John McCarthy, 'Ascribing Mental Qualities to Machines', manuscript, Computer Science Department, Stanford University, 1979 (emphasis in original).
2. Marvin Harris, *Cultural Anthropology* (Boston: Allyn & Bacon, 1999), pp. 23–24.
3. Ruth Millikan, *Language: A Biological Model* (Oxford: Clarendon Press, 2005).

Index

Asterisks (*) mean that relevant material is present only in a footnote; italic page numbers identify relevant Figures.